RISK-BASED METHODS FOR EQUIPMENT LIFE MANAGEMENT: AN APPLICATION HANDBOOK

A Step-by-Step Instruction Manual With Sample Applications

Prepared by

THE RISK-BASED INSPECTION APPLICATION
HANDBOOK RESEARCH TASK FORCE

Appointed by

THE PLANNING COMMITTEE FOR CODIFICATION
AND STANDARDIZATION RESEARCH* OF THE
ASME CENTER FOR RESEARCH AND TECHNOLOGY DEVELOPMENT

For

THE NATIONAL RURAL ELECTRIC COOPERATIVE ASSOCIATION
THE EMPIRE STATE ELECTRIC ENERGY RESEARCH CORPORATION
THE TENNESSEE VALLEY AUTHORITY
INDUSTRIAL RISK INSURERS
US COAST GUARD MARINE NATIONAL MARITIME CENTER

Reviewed and Edited by

THE STEERING COMMITTEE FOR
THE RISK-BASED INSPECTION APPLICATION HANDBOOK AND
AN INDEPENDENT PEER REVIEW COMMITTEE

ASME INTERNATIONAL
Three Park Avenue, New York, New York 10016-5990

* Succeeded by the Research Committee on Risk Technology.

DISCLAIMER

This report was prepared as an account of work sponsored through The American Society of Mechanical Engineers (The Society) Center for Research and Technology Development by:

Empire State Electric Energy Research Corporation
National Rural Electric Cooperative Association
Tennessee Valley Authority
Industrial Risk Insurers
US Coast Guard National Maritime Center

(Collectively referred herein as "the Sponsors").

Neither The Society nor the Sponsors nor the Industrial Risk Insurers, US Department of Energy, National Rural Electric Cooperative Association, Union Carbide Corporation and Southwest Research Institute (collectively referred herein as "the Sponsorees"), nor any financial contributors or others involved in preparation or review of the report, nor any of their respective employees, members, or persons acting on their behalf, makes any warranty, express or implied, or assumes any legal liability or responsibility for the accuracy, completeness, or usefulness of any information, apparatus, product, software or process disclosed, or represents that its use would not infringe privately owned rights.

Reference herein to any specific commercial product, software, process or service by trade name, trademark, manufacturer, or otherwise, does not necessarily constitute or imply its endorsement, recommendation, or favoring by The Society, the Sponsors, the Sponsorees, financial contributors or others involved in preparation or review of this report, or any agency thereof. The views and opinions of the authors, contributors, and reviewers of the report expressed herein do not necessarily state or reflect those of The Society, the Sponsors, the Sponsorees, financial contributors or others involved in preparation or review of this report, or any agency thereof.

Statement from By-Laws: The Society shall not be responsible for statements or opinions advanced in papers . . . or printed in its publications (7.1.3)

INFORMATION CONTAINED IN THIS WORK HAS BEEN OBTAINED BY THE AMERICAN SOCIETY OF MECHANICAL ENGINEERS FROM SOURCES BELIEVED TO BE RELIABLE. HOWEVER, NEITHER ASME NOR ITS AUTHORS OR EDITORS GUARANTEE THE ACCURACY OR COMPLETENESS OF ANY INFORMATION PUBLISHED IN THIS WORK. NEITHER ASME NOR ITS AUTHORS AND EDITORS SHALL BE RESPONSIBLE FOR ANY ERRORS, OMISSIONS, OR DAMAGES ARISING OUT OF THE USE OF THIS INFORMATION. THE WORK IS PUBLISHED WITH THE UNDERSTANDING THAT ASME AND ITS AUTHORS AND EDITORS ARE SUPPLYING INFORMATION BUT ARE NOT ATTEMPTING TO RENDER ENGINEERING OR OTHER PROFESSIONAL SERVICES. IF SUCH ENGINEERING OR PROFESSIONAL SERVICES ARE REQUIRED, THE ASSISTANCE OF AN APPROPRIATE PROFESSIONAL SHOULD BE SOUGHT.

Book No. I00474

ISBN 0-7918-3507-3

ACKNOWLEDGMENTS

This Handbook bridges the gap between a previously-laid theoretical foundation of risk-based methods and the practical needs of users in industry. Any bridge needs a firm footing; previous work by and contributions from many leaders in their respective fields from academia, government, and industry on risk-based methods provided a solid point of departure for this handbook.

In addition to its primary offering of procedures that provide access to risk-based methods, this Handbook makes available two unique resources, *The Fault Tree Handbook*, NUREG-0492, which is now out of print, and IRRAS v. 4.0, an early but powerful risk analysis tool. We are grateful to the US Nuclear Regulatory Commission, and particularly Mr. Jack Gutman and Ms. Marge Sheehan for allowing us to bring you these materials. Mr. Curtis L. Smith, INEL — Lockheed Martin, was most helpful with the IRRAS materials.

We are also grateful to G. Michael Curley, Manager-GADS Services at North American Electric Reliability Council, who supported development of this handbook by providing a copy of the pc-GAR computer program and the accompanying GADS database. We also appreciate his participation in numerous telephone conversations during which he clarified pc-GAR and database operations.

The steering committee members have carefully guided the project. The independent peer reviewers teamed with the steering committee to diligently review and edit this document. The valuable and generous contribution of these members, who are identified in this document, is most appreciated. All the steering committee members provided useful comments that were based upon informal reviews of the document. We are especially grateful to Tom Gaidry, Long Island Lighting (LILCO), George Montgomery, Progress Energy, Brett Shelton, Dominion Energy and Graham Cell, SWRI, who exercised the software and tested the procedures.

The research task force acknowledges with appreciation the unique contributions of Mr. O. J. "Vic" Chapman, Dr. David O. Harris, and R. Scott Hartley.

We are especially and eternally grateful to Mr. Raymond J. Art, recently retired assistant director of the ASME-CRTD, for his unwavering support and confidence throughout a project that seemed to last forever.

We acknowledge the ASME Technical Publishing Department members for their dedicated and diligent efforts toward compiling, editing, and publishing this document. We also acknowledge and thank the participants' families for their support and for the incredible patience and understanding that they have provided.

This Handbook is based upon the series, "Risk-based Inspection—Development of Guidelines." The series includes:

- Vol. 1 General Document also published by the U.S. Nuclear Regulatory Commission as NUREG/GR-0005, Vol. 1
- Vol. 2 Part 1 Light Water Reactor (LWR) Nuclear Power Plant Components also published by the U.S. Nuclear Regulatory Commission as NUREG/GR-0005, Vol. 2, Part 1
- Vol. 3 Fossil Fuel-Fired Electric Power Generating Station Applications

EXECUTIVE SUMMARY

The Handbook is a complete training tool for engineers who want to apply risk-based methods in an industrial setting. It is focused on senior engineers for program management and staff engineers for program implementation. Step-by step examples, complete with most of the specialized software that is needed to run them, use real-life input from fossil fuel-fired power plant experience to illustrate its techniques. The Handbook user must only provide a personal computer with standard business software and a statistical analysis program. Although the examples in the book are based upon fossil fuel-fired power plant equipment, engineers at any industrial facility can readily modify the examples and the demonstration software so that they will support a local risk-based program.

Risk-based programs are an extended framework and logical structure for decision making. They are probabilistic rather than deterministic in nature. Therefore, these methods acknowledge and systematically treat the uncertainty that actually underlies all component inspection planning and fitness-for-service decisions. Further, risk-based program models can identify the additional information that will most effectively reduce uncertainty. Finally, risk-based programs self-improve by accepting feedback from inspections, tests and analyses.

Previous documents in the ASME risk-based inspection series addressed an audience that was already familiar with probabilistic methods and with advanced risk-measurement concepts. Peer reviewers for Risk-Based Inspection — Development of Guidelines, Volume 3, Fossil Fuel-Fired Electric Power Generating Station Applications (ASME, 1994) suggested that we write this Handbook. They realized that risk-based methods application could greatly benefit smaller utilities, but that these utilities were not likely to have staff trained to implement the methods.

For many reasons, engineering analysis alone can no longer support

technical decision making. ASME, 1994, in addition to extending risk-based methods beyond the nuclear arena, provided economic analysis tools. These tools allow engineering analysts to calculate the long-term financial benefits that may be expected from a proposed strategy and to communicate these benefits to financial decision makers in a format that will be understood by those decision makers.

Risk-based programs offer significant value, because risk-based methods and strategies focus inspection and maintenance programs on the most economically effective areas in a way that meets safety and environmental risk standards and satisfies other user-selected criteria, such as reliability and availability requirements and budget limits. Therefore, risk-based programs optimally allocate limited inspection and maintenance resources and also provide rational economic support for safety-enhancing measures.

Further, risk-based programs provide management with the likely dollar value of prevented equipment failures. Feeding inspection results back to a risk-based program measurably increases confidence in facility integrity and reliability, even when no conditions are found to require correction. Risk-based programs can be evaluated for their likely effectiveness before they are applied.

Individual risk-based programs are living programs that are based upon rapidly-growing technology. The risk based programs and program development processes that are described in the Handbook are by necessity a limited subset of the whole available technology. The risk-based "toolkit" that we are offering includes:

- Qualitative assessment, for preliminary ranking.
- Quantitative assessment, using generic data, expert opinion elicitation and engineering models.
- System analysis, using fault tree and event tree analysis
- Optimized component replacement and major inspection management
- Inspection program development

Although we have assembled them in a somewhat unusual mix, we have only used well-established, generally classical engineering and fi-

nancial methods. By introducing the user to risk-based tools that range from the qualitative to the fully quantitative, we have tried to make available the greatest possible return. There are other ways to develop risk-based programs and other tools that could be applied to them.

The ASME encourages risk-based inspection and maintenance programs because they enhance safety. In some facilities, such as nuclear power plants, there is a severe safety consequence, like core damage, that overrides all other considerations. Fossil fuel-fired power plants have no such exposure. Equipment failures, no matter how serious, are unlikely to endanger life and property outside the facility. Most equipment failures and personnel injuries are associated with upset or abnormal operating conditions, which include startup and shutdown. Therefore, safety and reliability are closely related; safety improvements tend to improve reliability and reliability improvements tend to improve safety. Further, in addition to being safer, planned outages for maintenance are generally more economical than forced outages caused by failures.

Risk-based methods offer proven safety and economic benefits. For example, in the mid-1980s, the Niagara Mohawk Power Corporation developed a plant-specific failure database and used risk-based methods to support a life extension program for a 35-year-old fossil power plant. Their program focused on components that contributed the most to plant unavailability. When the study began, the overall plant availability was 70 percent. By 1989, plant availability had risen to 85 percent. Operator exposure to hazardous conditions was correspondingly reduced. At the same time, Carolina Power & Light used risk-based methods to optimize the replacement schedule for critical fossil fuel-fired power plant tube components. The resulting plan increased system net present value by more than $50 million over the value calculated for the original plan, which had been established using conventional engineering approaches. The Handbook offers a structured approach to decision making. Some decisions have political, emotional and other influences. When these influences are predictable, they can be modeled.

The bottom line:

- This Handbook guides its users through a step-by-step risk-based program development process.

- The process uses the tools of engineering, decision analysis and finance.
- The Handbook will help its users to produce safe, technically sound and economically supportable programs.
- It will show its users how to present those programs in terms of the value measures that senior managers have come to expect.

RESEARCH TASK FORCE ON RISK-BASED INSPECTION APPLICATION HANDBOOK

Michael E. G. Schmidt, PE, Research Consultant, Industrial Risk Insurers, Hartford, Connecticut, Principal Investigator

David A. Mauney, Senior Consultant, Structural Integrity Associates, Inc., Rockville, Maryland

Joseph Matrisoto, Mechanical Engineer, New Madrid Power Plant, New Madrid, Missouri

Lloyd G. Smith, PE, Project Manager, DOE-LAAO, Los Alamos, New Mexico

Joseph Balkey, Staff Engineer, Union Carbide Chemicals and Plastics, South Charleston, West Virginia

RESEARCH STEERING COMMITTEE ON RISK-BASED INSPECTION APPLICATION HANDBOOK

Raymond J. Art, Assistant Director, ASME Center for Research & Technology Development, Washington, DC (now retired)

William G. Wendland, PE, Manager — Nuclear Projects, American Nuclear Insurers, West Hartford, Connecticut

Dr. David O. Harris, Vice President and Principal Engineer, Engineering Mechanics Technology, Inc., San Jose, California.

Kenneth R. Balkey, PE, Fellow Engineer, Westinghouse Electric Company, Pittsburgh, Pennsylvania.

Zbigniew J. Karaszewski, President, Intertech, Ltd., Manassas, VA

Dr. Robert Perdue, Westinghouse Electric Corp., Pittsburgh, PA

Prof. Bilal Ayyub, University of Maryland, College Park, MD

O. J. V. Chapman, Rolls Royce & Assoc., Ltd., Derby, United Kingdom

INDEPENDENT PEER REVIEW COMMITTEE

Prof. Dan Crowl, Michigan Technological University, Houghton, Michigan

Michael Curley, Manager — GADS Services, North American Electric Reliability Council, Princeton, New Jersey

Dennis Hendershot, Rohm and Haas Company, Bristol, Pennsylvania

David Moser, US Army Corps of Engineers, Alexandria, Virginia

George Montgomery, Carolina Power and Light, Raleigh, North Carolina

For Appendix D:

David Gertman, INEL/Lockheed Martin Idaho Technologies Company, Idaho Falls, Idaho

CONTENTS

CONTENTS

CONTENTS

SECTION 1

This is not a "read" book, it's a "work" book. If you've read this far in this book, you almost certainly have a problem. On the other hand, if your facility is new and you have unlimited maintenance resources, or maybe it's not so new but it has always had unlimited maintenance resources, if your corporate culture explicitly embraces the long view and a proactive stance, if maintenance people are just as much heroes as production people (all the time, not just when they respond effectively to a breakdown) — well, you don't need this book yet. You might read it anyway; you might get some good ideas for now, and when the future forces hard choices upon you, because of equipment aging, shifting markets and competition or management changes, you will be ready with a structured way to make those choices.

Throughout the Handbook, the page that faces the first page in each section will be an outline map that lays out the section prerequisites and expected outcomes. The map for the introduction follows.

This step requires:

- *A specific interest in risk-based methods, or*
- *A stake in maintenance decision making and a need for a structured way to make such decisions, or*
- *An environment that "experience-based" or strictly engineering-oriented methods alone cannot proactively address.*

This step:

- *Introduces probabilistic and risk-based methods.*
- *Argues for a financial basis in engineering decision analysis.*
- *Discusses how to use the Handbook.*
- *Previews the specific risk-based toolset that the Handbook will present.*

The next step will, in general, be program development.

INTRODUCTION 1

Equipment begins to age as soon as it is built. Cyclic stresses cause looseness and fatigue. High temperatures cause creep. Erosion and corrosion thin and weaken. Age in many ways degrades, deteriorates and destroys.

Maintenance manages the aging process. Inspections and tests detect deterioration. Repair and restoration limit damage or restore the damage level to that which originally existed. However, different components age in different ways and at different rates. Different component failures have different consequences.

The most obvious way to enjoy a safe and failure-free existence is to find the component that ages the fastest, establish an inspection and replacement program that ensures that component's integrity, and then inspect and replace all the remaining components the same way. It is not difficult to guess that such a program will fail, because it will bankrupt its practitioners. What is really needed is a decision-making structure that acknowledges and accounts for the following:

- Similar component groups, in some cases similar components, often have different failure consequences. A waterwall tube failure in a 600 MW unit is probably more serious than a waterwall tube failure in a 100 MW unit.
- Similar components or component groups may require different maintenance strategies. In a power boiler, superheater and economizer tube ruptures have similar consequences, but superheater and economizer tubes rupture with different frequencies for different reasons.
- Damage accumulates in a manner that is neither constant nor linear over time, particularly later in life. For a complex component or system or for a large fleet of components, failure rate follows a

"bathtub" curve. Early failures, which generally involve design errors or manufacturing defects, provide an initially high but decreasing rate ("infant mortality"). This "break-in" period is followed by a period of essentially constant failure rate, during which damage that is caused by long-term failure modes accumulates. This damage dominates during the final period ("wearout"), which is characterized by a rising and accelerating failure rate. (See Figure 1.1)

- The bathtub curve reveals a few lessons and cautions. Although the falling failure rate tends to suggest that "things are getting better," they are not. The cumulative failure probability, which tracks the total system exposure to harm, is a function of the area under the failure rate curve and is always increasing, though not always at the same rate. That attractive drop in the failure rate suggests that certain failure modes will run their course quickly or will not develop at all. A "warranty inspection," which is a thorough inspection that is typically conducted shortly after commissioning, seeks to detect such failure modes. If no such failure modes are found, then the longer-term failure modes become the proper focus and scheduling basis for subsequent inspections.
- An aggregate component, such as a complete superheater, will reach a point in its life when completely replacing it is more efficient than replacing its individual elements as they fail.

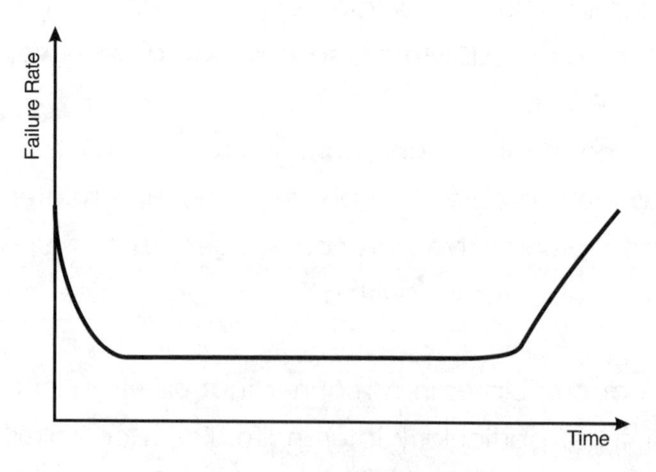

Figure 1.1 "Bathtub" Curve

- Both the "best case" scenario and the "worst case" scenario are rare events. Although the worst case needs to be considered for contingency planning reasons, the worst case is not, in general, a sound economic basis on which to build a maintenance program.

Risk-based analysis methods integrate available knowledge and experience about your facility and also certain key safety criteria into a rational framework. That framework will ensure that your safety criteria are met while it helps you to optimally schedule and plan major maintenance activities. Risk-based analysis uses standard financial tools to justify those maintenance activities in a way that can be easily communicated to and understood by corporate staff and management.

This Handbook provides the knowledge and resources that you will need to start building a risk-based inspection and maintenance program at your facility.

- It starts with qualitative methods that risk-rank components and set priorities.
- It continues with more elaborate risk-ranking methods that use generic data, your own data, expert elicitation and engineering models to develop a quantitative risk ranking and predictive risk measures.
- It shows you how to use the supporting models, various financial information, budget limits and safety, environmental and other constraints to identify the most important components and determine when they should receive attention.
- It concludes with formal decision analysis methods that tie all the data into optimized component life management programs. These programs might include:
    - ◇ Routine and special inspection schedules and strategies
    - ◇ Condition monitoring and remaining life determination
    - ◇ Optimal component repair and replacement schedules.

Every program development step adds value. This means you do not have to develop a fleet-wide inspection program before you get useful results. Further, the program development process and the programs themselves include many feedback opportunities. As you will see, a risk-based program constantly improves as you add more information.

All the quantitative programs that you develop using this Handbook can be constrained by explicitly defined limits for safety, forced outage rate (on-line availability) and other user-selected factors. All optimizing calculations, while bound by user-selected limits, maximize the expected net present value. This common financial basis allows you to compare options; more important, it gives you results that are effectively communicated to and understood by facility management. Basing the safety criterion upon failure probability and then optimizing based upon economic return ensures that the safety goals are met while it ensures the highest economic return. This procedure avoids the ethically and legally difficult problem of trying to price life-risk that results from various hazards.

You will note that the procedures up through qualitative risk assessment involve little more than organizing and tabulating facility data and records. This statement does not intend to trivialize organizing and tabulating records and qualitative methods; these activities can require a major investment. You may, by performing these procedures, "harvest enough low-hanging fruit," that is, resolve enough fairly easy issues, to build support for the more difficult quantitative procedures. You may also be satisfied to stop at the qualitative level, though if you do, you will not be able to use the available financial tools.

The Handbook provides only a sample of the risk-based methods that exist. Because this is an introductory "how-to" book, we could only select and offer a single path from deterministic methods to fully quantitative risk-based methods. Other paths exist. We have restricted ourselves to well-tested, mostly classical engineering, financial and mathematical tools, though we have sometimes combined them in unusual ways.

Do not be intimidated by the apparent scope and complexity of what's offered here, nor by the use of techniques that engineers seldom use. If you don't modify the procedures at all, you will still obtain sound results in a format that most managers will positively receive. If you are working wholly on your own, you should not initially replace the default values in the procedures that you perform during the learning process; you should merely note what areas might need refining later for specific applications.

You may find that some information, like specific component failure rates, is not readily available. You may surprise people who do not expect engineers to be looking for other information, such as projected capacity factors, replacement power costs or the corporate value measure. Keep in mind, however, that the procedures will work acceptably with generic data and default values. Having to modify and rework the procedures is to be expected. A risk-based program is an iterative, or living process that can always be improved.

This Handbook supplements the ASME research series "Risk-Based Inspection — Development of Guidelines." "Volume 1 — General Document" (ASME, 1991) introduces the risk-based process. "Volume 3 — Fossil Fuel-Fired Electric Power Generating Station Applications" (ASME, 1994) applies risk-based methods to fossil fuel-fired electric power generating station component inspections. Volumes 1 and 3, however, supply broad concepts that must be tailored to individual applications. This Handbook provides step-by-step instructions that you can use to apply risk-based concepts to your facility. Although the examples in the book are based upon fossil fuel-fired power plant equipment, you can apply the risk-based concepts to any industrial facility.

You don't need Volumes 1 and 3 to start using the Handbook. As your risk-based programs develop, you'll find the Handbook's limits and want more information. Volumes 1 and 3 contain more information, and they also reference other resources. If you have questions or concerns along the way, we encourage you to contact the ASME Research Committee on Risk Technology for assistance. You can contact the committee through the ASME Center for Research and Technology Development. See Appendix F for more information.

1.1 □ RISK-BASED INSPECTION METHODOLOGY

This section introduces terms that the later procedures use frequently. Underlined words are defined or discussed in the immediately surrounding text. **Boldfaced** words are defined or are further discussed in Section 13. Appendix C.1, The Fault Tree Handbook (FTHB), which is located on the Handbook CD-ROM, also introduces many useful risk analysis concepts. The FTHB also contains a much more detailed introduction to statistics and probability. Its presentation is slightly more formal than that which follows, and it includes mathematical theory and illustrations.

1.1.1 □ Risk and Probability

Risk is a measure that combines failure **probability** and failure **consequence.** Failures can have many sorts of consequences. The procedures in this Handbook generally keep the safety issues on the probability side of the ledger (the Executive Summary explained why) and therefore deal with only economic costs on the consequence side. To emphasize this point, the rest of the Handbook refers to "consequence cost" instead of "consequence" unless the topic explicitly considers other sorts of consequences (e.g., in qualitative risk assessment).

The quantitative risk ranking procedures and examples in this Handbook calculate risk by multiplying probability and consequence cost, however, risk is not necessarily such a straightforward function. Although risk itself may be mathematically precise, risk tolerance is not. You must beware of calculated risk values that conceal unacceptably high probabilities (unacceptable nuisances) or unacceptable consequences.

Certain values, such as operating temperatures and various material properties, will enter into calculations as **probability distributions.** There are many kinds of probability distributions, however, those which are most useful in risk-based calculations "peak" at the most likely value and are "shaped" in a way that reflects the variability of the value and our **uncertainty** or level of confidence in that value. Variability is a data characteristic that you may or may not be able to control. For example, you probably cannot improve a failure probability estimate for a complex component like a boiler feed pump unless you consider factors such as age, history, present vibration level, etc. Although failure probability and failure consequence cost will in general be uncertain, in this volume we will use the mean or average value to represent the most likely probability or consequence. In other words, we will address neither the uncertainty of the uncertainty nor the uncertainty of the failure probability

The risk-based inspection methodology that is used in this handbook is therefore **probabilistic** rather than **deterministic.** This is because the reality that is being **modeled** is too complex and uncertain to be modeled in sufficient detail for a deterministic solution. Further, failure probability and failure consequence cost generally vary with time. Failure probability increases because of aging or accumulated damage, and decreases because of component repair or replacement. Failure consequence cost is sensitive to a variety of financial factors.

There are two kinds of probabilities. Frequency-related probability is the more common type. It is associated, for example, with the number of times in a given number of trials that a flipped coin will come up heads or the frequency with which a fair pair of dice will produce a given number. The other kind of probability is belief-related or subjective probability. It is associated, for example, with a weather forecaster expressing the chance of rain tomorrow.

The probabilities that we derive from failure history data are frequency-related because the probability of each failure event is the fraction of the total failure events that occur to all the applicable systems or components during the period that the data history covers plus those that are still operational. Our data is never "complete," however, because some or all of the systems or components are still in service when we calculate failure probabilities. This is called right censoring. We must account for the systems or components that are still operational as well as those which have failed, because if we do not, we would assume a 1.0 failure probability at the end of the period of interest. The adjustment also accounts for the fact that failed systems or components are generally repaired or replaced and returned to service.

We obtain subjective probabilities when we interview experts about when a failure might occur. We also obtain subjective probabilities when we use engineering models to estimate remaining life. Even though we use a numerical analysis to estimate remaining life, the result is still just a belief that the component or system will fail at or by a specific time.

There are two ways to express the probability of a failure occurring over time. The first is the probability that an event will occur at a specific time. For example, we might calculate the specific probability that a failure will occur during each of the next ten years. The resulting set of probabilities would define a probability density function or PDF. The

PDF expresses our "belief density" that failure will occur in each specific year. The second way to express the probability of a failure occurring over time is to consider whether a failure will occur in *or before* a certain year. A set of probabilities for failure occurring in or before a number of years would define a cumulative probability function or CDF. Each point on a CDF expresses our "belief density" that failure will occur in *or before* each year. See the Fault Tree Handbook, Sections 10.3 and 10.4, for more information about CDFs and PDFs.

Risk-based inspection methods generally use subjective probability. Subjective failure probabilities are based upon analysis of incomplete and uncertain information. Incompleteness may be reduced by research and analysis and uncertainty may be reduced by considering inspection results, but in a complex system, the failure probability will never be as concrete as the probability that a fair pair of dice will produce a seven one-sixth of the time that it is thrown. Subjective probability analysis is still useful; in fact, it is the only practical tool that will reduce complex physical phenomena to a form that is usable by a **decision analysis** process.

Note also that subjective probabilities are calculated values that can change without any change having occurred in the physical system to which they are applied. For example, results of an inspection that finds no adverse conditions, when they are fed back into a risk-based analysis, may produce a lower expected value for the failure probability, or improve the shape of the probability distribution. In other words, an inspection can reduce our perception of (or belief in) the uncertainty and/or variability. We might also say that inspections increase our **confidence** that our model accurately reflects the equipment condition. Understand, however, that confidence and success are not necessarily the same. A clean inspection might make you confident about success, but another inspection might make you confident that a key component is dying fast.

This is analogous to your weather forecaster altering his belief about his predicted rain probability upon receipt of additional information, for example, about an approaching cold front. The weather forecaster has obviously not changed the probability of rain, but has only provided a refined probability estimate. And neither your physical system nor the weather necessarily cooperate. Probabilistic analysis points you toward the most likely outcome. It certainly does not guarantee that the most likely outcome, or a "better" one will occur. Recall the casino; even with well-understood probabilities, "hot" or "cold" dice can give the casino dealer and customers rather exciting times over the short run.

In the preceding paragraphs, we have used terms like "perception" and "belief." This is to emphasize that the probability that we will be using is subjective. Recall the dice. If you rolled a pair of dice one hundred times and never rolled a seven, you would conclude that your model was in error, e.g., that you were not dealing with a fair pair of dice. A problem that you believed could be modeled would have become subjective. Likewise, no matter how well you think that you understand a system, you must always beware of the unknown. If your system consistently fails before inspections, you should be looking for failure modes or mechanisms or other factors that your model had not considered.

Uncertainty is a measure of our confidence in the expected value. Uncertainty arises from incomplete knowledge of the system that we are analyzing. Mathematically, uncertainty produces a "wide" or "flat" probability distribution. See Appendix C.1 Figure 10-5. However, **variability** can also produce a flat probability distribution. Variability is a measure of the inherent sloppiness of the real world. Although uncertainty and variability may look alike to a modeler, uncertainty is an attribute of our model, of our level of knowledge, or of the quality of our measurements of the physical system. Variability is an attribute of the physical system itself.

We can decide whether we need to worry about an uncertain variable that has a "bad" distribution by using **sensitivity analysis.** Sensitivity analysis is a tool that decision analysts use to identify which variables control the conclusion. Highly uncertain variables that sensitivity analysis identifies as important become candidates for further study and analysis. Decision analysis tools can also determine the value of additional information. More information will not help, however, if the highly uncertain variable is inherently uncertain, e.g., has high variability. For example, if local carbon content in a steel structure is very important to us, but the quality control in the mill that produced the steel was so bad that the carbon content is "all over the map," then no amount of research is going to buy us a "better" distribution. The most accurate possible model is one in which the uncertainty and variability match. Beyond this point, no further improvement is possible.

We have spent a lot of time discussing this point because it is one of the "hard" limits on the power of risk-based analysis methods. If we cannot adequately contain the variability in a risk analysis, then we will have to take steps to reduce the uncertainty so that the risk analysis appropriately represents the situation, or we will have to consider other risk management tools, such as process or equipment redesign or protective systems.

Risk-based methodology ranks or groups **systems** and **components** by their **risk importance. Qualitative risk analysis** ranks systems and components relative to each other. **Quantitative risk analysis** measures or estimates risk numerically. However they are measured, the risk values are the key inputs to a decision-making process. In this handbook, qualitative risk values are used for component ranking or screening. Quantitative risk values are used in financial calculations that more precisely rank components and which can also identify the best time to inspect or replace them.

Quantitative risk values can help develop the **strategy** (i.e., the timing, method, and sample sizes) for performing **inspections.** Risk-based methodology focuses inspections where they are most needed and cost-effectively allocates inspection resources.

Human factors can be analyzed concurrently with physical factors. Human factors analysis can identify areas where better task design, operating practice and training will reduce risk. See Appendix D for more information about human factors analysis.

1.1.2 □ History

Formal risk-based methods, which are based upon several classical mathematical and engineering tools, recently came to prominence in the nuclear power industry. Researchers in that industry found that experience-based methods tended to ignore structural failures, such as pressure vessel and piping system ruptures, because these failures seldom occurred. Even when probabilistic methods began to appear, structural failures tended to be "sorted out" of programs because those failures were generally at least an order of magnitude less likely than active component failures, such as pump and valve failures. The risk basis, because it factored in the potentially severe consequences that structural component failures could produce, "leveled the field."

Most other industries, including fossil fuel-fired power generation, have structural components, such as boiler tubes, that have fairly well-recognized deterioration mechanisms and which fail much more regularly than the structural components in a nuclear facility. Other structural components, such as high-energy piping, drums, headers, turbine casings and generator retaining rings, have recognized deterioration mechanisms, but (fortunately!) lack failure experience. You can safely operate such components with far greater confidence under a risk-based inspection regime. A risk-based program can help manage any fleet of components in a cost-effective way.

There is one more significant difference between the typical nuclear facility and most fossil fuel-fired power plants and other industrial facilities. That is aging. Most nuclear facilities are in the constant failure rate regime that is shown in Figure 1.1. Most other facilities, or at least certain critical systems in those facilities, whether because of more severe environments, less strict design and construction specifications or absolute age, have entered the "wearout" regime. This fact greatly complicates maintenance calculations, especially for major (read "expensive") components. The accelerating failure rate that is a characteristic of wearout makes optimum repair and replacement strategy calculations mathematically complex. Various economic factors dictate that virtually all facilities, nuclear and otherwise, will be operated well into their wearout regimes.

1.1.3 □ Finance-Based Analysis

The quantitative analysis methods in this Handbook optimize decisions by maximizing the expected **net present value (NPV)** of the system being studied. This optimizing is done in a manner that ensures safety and other critical criteria are met. This financial basis might not be very popular with some engineers; however, no value measure is as versatile a tool as cash value for comparing different choices. Further, top management is ever less likely to accept anything else as the basis for making a decision about a proposed action plan.

There is a lot of emphasis today on "value added." This is because most companies are stockholder-owned. The companies need to maximize their return on invested dollars, especially in the area of maintenance. Maintenance costs receive particular scrutiny because maintenance is not often perceived as producing product or adding value. Using NPV as a financial decision-making value measure will help you achieve this "value added" objective. NPV, which is described in more detail in Section 10.1.4, is the tool you will use to show that maintenance adds value.

Maintenance people have often used engineering rather than financial analysis to explain and justify proposed activities. However, engineering-based proposals are difficult to weigh against financially analyzed proposals. This difficulty can cause maintenance needs to have trouble competing for attention. NPV is a value measure that all corporate groups can use when they communicate with the corporate decision-maker. Major corporate finance textbooks (Brealey, 1991) and (Brigham, 1991) recommend NPV, and the Handbook procedure and examples illustrate its use in the decision-making process. When maintenance needs and engineering facts are converted into cash flow and NPV terms, then they can compete on a level playing field with other corporate needs. Dollars are the language of the decision-makers and the stockholders. See Appendix A for more discussion.

Maintenance expenditures are investments in components that will operate for a number of years. Looking at the financial effects over time is more realistic than looking at initial cost alone. Therefore, the procedures in this Handbook not only consider the initial inspection or component replacement cost but also determine the expected cash flow consequences over the service life of the equipment being analyzed.

The "benefit" associated with the maintenance "cost" will be avoided production losses during the near, intermediate and long term.

Several of the procedures are optimizers, that is, they find the strategy that will maximize the expected NPV consistent with meeting safety and other goals. Optimization demonstrates probable savings by looking at long-range component life consequences rather than focusing on the immediate cost. Therefore, NPV is more than a communications tool. It is also a very useful parameter upon which to base our decision calculations because it measures their effects over the expected service life of our facility.

Some analysts base their project decisions upon payback periods. The difficulty is that payback analysis only looks at the short range. Looking at benefits, and any repeating costs, over the component service life provides a more realistic view. The analysis procedures in the Handbook can be arranged to maximize NPV over any interval. Effectively, the procedures calculate realistic values for long-range benefits by implementing the proposed plan one year at a time. This one-year-at-a-time optimization accounts for short-term changes and effects that need to be accounted for in any living decision process.

Using NPV as the value measure is a useful and widely accepted strategy that supports both short and long range views. For that reason, the Handbook procedures and examples will use NPV. If you must use another value measure, you can modify the financial portion of the appropriate spreadsheets to accommodate it.

1.1.4 □ Decision Analysis

Decision analysis is a science whose development was motivated by the complexities of World War II logistic and other problems. Over the last thirty years, decision analysis tools have been developed to model the process that connects the practical decisions of business, the inputs associated with these decisions and the financial yardstick that a company uses to measure success. The decision analysis process is meant to reflect the decision process as it is or is desired.

Effective engineering analyses need to feed into the corporate decision making process. Because engineering and financial analysis are so different, a modeling process that can bridge the gap between them is necessary. This is especially true in the area of maintenance, where the link between engineering and corporate financial significance is less clear. Decision analysis bridges that gap.

Financially based investment decisions usually seek to create the highest NPV. Engineers have often used "worst-case" scenarios to address uncertainty. Decision analysis, which is often used by financial analysts, accounts for uncertainty directly instead of using worst-case analysis to address it indirectly. The "worst-case" approach "builds in" excess conservatism that today's competitive environment can no longer afford. Further, the worst case is just as rare an event as the best case, therefore, neither is a rational basis for decision making. A principle of decision analysis states that decisions that are based on expected values rather than on the worst cases are more likely to gain than to lose. This Handbook will therefore follow the growing business trend and use expected values for decision making instead of just the worst cases.

The worst cases still need to be considered, because these cases might represent unacceptable losses. A possible loss is "unacceptable" if management is not willing to live with its calculated risk. Losses greater than 15% of the total corporate asset value and large losses that involve public consequences are generally unacceptable. In such cases, risk aversion must be included in the analysis, or risk management strategies must be considered. Both options are beyond the scope of this Handbook, but the techniques can be found in the decision analysis references (Section 15).

Decision analysis mathematically models processes. However, initial model construction and later concept communication is better done graphically. Influence diagrams were invented to graphically present decision models. During model construction, influence diagrams allow decision analysts to "brainstorm" relationships without getting mired in mathematics. After the analysts have defined the relationships and obtained the needed inputs, the influence diagram will help them to construct the mathematical model that will perform the calculations. Later, influence diagrams illustrate the relationships among the decision model elements separately from their mathematical details.

1.2 □ OBJECTIVES AND SCOPE

This Handbook introduces risk-based methods. It is also a step-by-step program development manual that will help you to immediately apply those methods.

We assume that you are an engineer in an industrial setting and that you have questions like these:

- "How do I organize the accumulated information that resides in operating logs and in outage, failure investigation and inspection reports? And how can I use this information after it's organized?"
- "I've been assigned to reduce forced outages for our most efficient unit. What component, system or procedure do I study first?"
- "Management has cut the maintenance budget. Can I safely achieve the required unit forced outage rate limit with the available resources? If so, how? And if not, how do I justify more resources?"
- "I learned about a better nondestructive examination (NDE) method for boiler tubes. Should I use it?"
- "My company has four units over thirty years old. Major components have dwindling remaining lives. In what order and when do I replace them?"

The Handbook will guide you to answers to questions like these. It explains how you should organize and analyze your data, how you might supplement that data, and how to establish a complete risk-based inspection program. The Handbook is liberally illustrated with examples that are based upon fossil power plant generating equipment. These examples are ready to be used as-is; you only need to substitute local data. Aside from basic engineering knowledge, you only need to provide information about your facility and its equipment and a personal computer with standard business software and a statistical analysis package. See Appendix B.

The Handbook is designed to obtain "buy-in" to risk-based technology. It does this for you by giving you a powerful tool to help you do your job. It also does this by offering paybacks at every stage. You can present these paybacks to management as evidence of success that can support program continuation and expansion. "Bottom line" successes that have already been obtained by persons using risk-based methods are noted in the Executive Summary.

The risk-based inspection methods presented in this document have been tested. This Handbook presents methods, some of which are unique, that quantitatively measure and compare the risk importance of individual components across various plant systems and among various units and facilities. Although plant availability, forced outage rate and property damage are the significant risk measures in fossil plant equipment methods (see the Executive Summary for an explanation), you can include explicit safety criteria. You can also introduce environmental and other constraints.

Be careful when you apply risk-based methods to inspection programs for major components, such as boilers, on which jurisdictional inspections are currently performed. Activities that routinely occur during these inspections, which are generally conducted annually, powerfully influence component availability by arresting developing conditions that could lead to failure. Unless you expend significant effort to capture complete information on all the repairs that are made during outages, you may obtain a distorted picture of the risks inherent in this equipment. You could, as a result, base an attempt to relax the inspection requirements on false data. In any case, jurisdictional requirements override any conclusion that may be reached by a risk-based program. You can, however, use the "jurisdictional effect" to your advantage. You could include the jurisdictional inspection as a fixed point in your inspection strategy, therefore simplifying its development, or you could use your program results to build a case with the jurisdictional authorities to possibly obtain permission for an alternate jurisdictional inspection strategy. (In fact, deciding which option is better is an excellent, if somewhat advanced, risk-based program application.)

1.3 □ USING THE HANDBOOK

1.3.1 □ Handbook Contents

The Handbook has several parts, all of which will, in different ways, help you to understand and apply risk-based methods:

- The Handbook itself; the printed book you are now reading. It explains risk-based methods in a conventional "textbook-like" manner and tells you how to apply them. It ties all the other parts together. You would probably do well to read at least Section 1 of this book first, to get the "big picture." This book refers you to the tools that are provided on the CD-ROM.
- A file called RBMAP.PDF. This file, which is accessed through Adobe Acrobat Reader, is a detailed, interactive flowchart that links to the step-by-step procedures.
- The CD-ROM \EXAMPLES subdirectory, which contains templates, or example spreadsheets for most of the handbook procedures. You can copy these examples and exercise them until you understand them thoroughly. Then you can strip out the example data and replace it with your own data and modify the basic spreadsheets so that they apply to your facility. Read Appendix B before you run any of the Handbook spreadsheets.
- Other CD-ROM subdirectories that contain the NERC-GADS cause codes (Appendix E), reference materials on fault tree analysis [the Fault Tree Handbook (FTHB)], the Integrated Risk and Reliability Analysis System software (IRRAS) and its tutorial and reference manual, and Adobe Acrobat Reader. See Appendix C.

The Handbook follows this format:

SECTION 1.4 outlines the risk-based inspection development process that is presented in the balance of the Handbook. The process is a series of modular activities.

SECTION 2 briefly explains the purpose of each activity and identifies the prerequisites to the activity and the product(s) of the activity.

SECTIONS 3–12 provide, for each activity:

- A more detailed discussion of its purpose
- Any needed theoretical background

- Step-by-step instructions
- One or more examples

SECTION 13, DEFINITIONS, provides additional commentary about some of the terms that the Handbook uses. Recall that the FTHB more rigorously defines many of the terms, mostly those that involve probability and statistics.

SECTION 14, CLOSING COMMENTS, offers a few procedural cautions and also identifies certain technological and other weaknesses in the Handbook.

SECTION 15, REFERENCES, is a short bibliography of risk-based methods resources and a short list of possibly useful software. Neither list is exhaustive.

APPENDIX A, WHY FINANCIAL METHODS; a brief discussion that explains and supports finance-based methods;

APPENDIX B, COMPUTER AND SOFTWARE REQUIREMENTS, which discusses the computer resources that you will need;

APPENDIX C, TOOLS, which contains instructions for accessing The Fault Tree Handbook and IRRAS, its tutorial and reference manual;

APPENDIX D, HUMAN FACTORS, a discussion of human factors analysis, complete with examples;

APPENDIX E, INDEX TO NERC-GADS COMPONENT CAUSE CODES;

APPENDIX F, HOW TO GET HELP

The Handbook uses the following conventions:

- <Enter> is a key on your computer keyboard.
- <Cntrl-C> is two or more keys on your computer keyboard that are pressed simultaneously.
- {SYSTEM} is a database field or a spreadsheet column.
- [Calculate] is a "click box," menu item or other mouse-selectable action.
- [Edit] [Paste Special] [Value] is a multilevel selection, e.g., in the Edit menu, select Paste Special, in the box that appears, select Value.
- \SUBDIRECTORY is a Handbook CD-ROM subdirectory.

- NAME.EXT is a Handbook CD-ROM file. (We restricted ourselves to the "8.3" naming convention to eliminate the problems that long file names sometimes cause. If your system allows long file names, by all means use them when you copy, rework and rename the examples.)

1.3.2 □ Handbook Usage Hints

You might develop a risk-based inspection program this way:

1. Read the Handbook; test the software[3]; see and understand what risk-based methods can do for you.
2. Determine what data may be available locally, and begin gathering it.
3. Apply the software that is provided in the Handbook to your data; begin modifying the routines to suit local conditions and extend the routines to other applications.

The Handbook is designed to get you started as quickly as possible. The example spreadsheet templates on the Handbook CD-ROM are already filled with sample data. When you read about a spreadsheet procedure, put down the book, fire up your computer, load the sample spreadsheet and work through the examples while you read about them. We deliberately included only enough illustrations to give you feedback (like you opened the correct spreadsheet) and not enough to follow the procedures without walking through them yourself.

The Handbook CD-ROM has a significant benefit inherent to its format: you cannot overwrite it. Therefore, copy the file you want to try out, walk through (well, click through) the procedure once with the preinstalled data and default settings and understand how it works. Then, freely make the procedure do your bidding, secure in the knowledge that you can start over if you err terminally.

You are now without doubt ready to plunge eagerly into risk-based methods applications. Before you do so, please consider the following:

[3]You should not, at this point, attempt to apply the software, rather, you should, for lack of a better word, play with the software. This, by the way, is a highly under-rated technique that can sometimes substantially reduce the slope of the learning curve for new software.

1. Your program could become a significant corporate asset. Build it accordingly.
2. Start with a program directory and subdirectory structure on your hard drive that is dedicated to the project. You'll need structure sooner or later, so start off on the right foot.
3. Freely but meaningfully rename the spreadsheets as you modify them. Not only does this habit give you some idea what each spreadsheet is supposed to do at some time after you have modified it (this is an iterative process!), the habit reduces the amount of work you must repeat if you introduce a fatal flaw. If the "last saved version" is the one on the CD-ROM, recovery from such a flaw could be time consuming.
4. Annotate. Possibly the second most important thing you can do. You will be modifying procedures, making assumptions, playing "what-if" games, seeking an optimal solution to a complex problem — you will do yourself no favor if you can't remember what you did from one loop to the next. You can, for example, use the "cell note" feature in Excel.
5. Backup. We saved the best for last. See item one of this list and the Executive Summary (which is modest in its estimates of the value that you could possibly contribute) for possibly millions of reasons why you don't want your program to die with a damaged hard drive.

1.4 □ PROGRAM DEVELOPMENT OVERVIEW

A risk-based inspection program could implement all the process modules that Figure 1.2 illustrates and more. You don't need to immediately commit the resources that such a program would require, however, each module produces value and each step justifies the effort that you expend upon it. Further, your risk-based program need not apply initially to an entire facility, unit or piece of equipment. Developing a risk-based inspection program for the tubes in a particular boiler would be a reasonable and useful way to start a risk-based program. You can expand the program as it proves itself. Your own creativity is your only limit.

Your inspection program will probably grow in two ways: It will grow

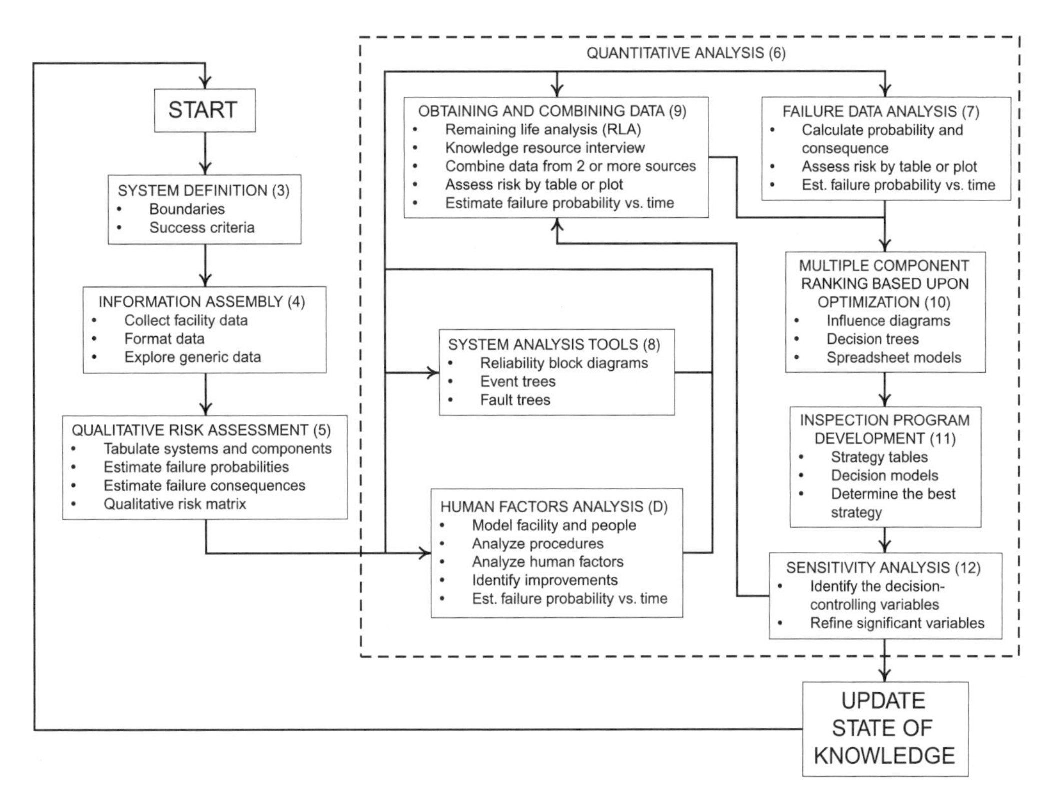

Figure 1.2 Risk-Based Inspection Process Overview

in depth and sophistication as you develop additional information or as you may need to reduce uncertainty. It will grow in scope for several reasons: The first issue that you investigate may have unsuspected roots. Or, when you address the most immediate issue(s), you will uncover other issues. Also, initial success encourages continued program development. Remember: the broader the scope, the greater the value return for your company.

1.4.1 □ SYSTEM DEFINITION

The first step in risk-based program development is system definition. A **system** is an identifiable entity that has an intended production function. More formally, a system definition has two parts. The system boundary identifies what the system is and what it is not. The success criteria identify what the system is expected to do. If you are installing a corporate-level risk-based program, and have been given the needed resources, you could define your system as your organization's total production assets.

However, even in this (unlikely) circumstance, you should probably start with a limited system definition.

Systems are made up of components. You will need to identify all the components of a system, and also identify how each component influences system performance. If you can't identify components, then the system definition was too narrow. For example, you have a boiler that is causing excessive forced unit outages. You could define the boiler as the system rather than the whole unit. Why is the forced outage rate unacceptable? Maybe because of tube failures. What kind of tubes? Maybe finishing superheater tubes, finishing reheater tubes and waterwall tubes. If you define the system as the boiler and then you focus on the components of concern, you are more likely to reduce your resource requirements to a manageable level. Systems, therefore, should be defined in the most exclusive way that provides a complete collection of critical components. This is particularly true if you are setting up a risk-based program on your own in your spare time. Success with a manageable issue is the event most likely to obtain management support for your program.

You define success criteria by using the same logic that you used when you defined the system boundary. When you start to develop a risk-based program, you should define success narrowly, along the lines of "acceptable forced outage rate between statutory inspections." Properly selected and well-defined success criteria will maximize the return on corporate assets.

System definition requires information. Sizes, shapes, materials, operating parameters, history; all must be known, measured or estimated. The amount of difficulty this presents depends upon the extent of the system boundaries and the quality of the available records. If your facility's prints are out-of-date, if important materials are of uncertain pedigree or if operating logs aren't being properly reviewed, just collecting the data you need to define your system, e.g., to "find out what's really there," will add value. Further, just collecting and organizing complete data will resolve some issues. For example, you might find a correlation between tube ruptures and cold starts. To ensure that the correlation is real and to determine its nature, you would first study the problem in greater depth. If you know the cause of the correlation, you can properly focus your corrective action. Learning that you have local overheating might lead you to alter your operating procedures. Learning that the failures also involve internal deposits might lead you to an entirely different response.

The data-organizing requirements and the focus on value-producing data are important "side benefits" of the risk-based process. Unfortunately, you might learn that, because of housecleaning or some less drastic reason, your data cannot support a risk-based program. Or, you might have access to a limited amount of data or data that covers only a single unit or a few units. In any case, if your facility-specific data are not sufficiently robust, you can use generic data to get your program started.

1.4.2 □ Qualitative Risk Assessment

Qualitative risk assessment is the next step. This step assumes that you have more than one "problem" or that your "problem" has more than one possible "solution," and you need an importance ranking. You can have more than one solution even with a narrowly defined system. If you're analyzing finishing superheater tubes, for example, you should find that some tube components are more important than others. If you don't, then you may need to investigate the problem further and describe the system in greater detail.

To perform a qualitative risk assessment, you might need to identify failure modes for each identified component. A failure mode is a defined way in which a component can fail to satisfy its success criteria. You will estimate the failure probability for each component or component and failure mode combination. For this step, selecting "high," "medium" or "low" occurrence probability is sufficient. For example, "reheater tube fails" might have a high probability and "condenser tube fails" might have a medium probability. In a component-and-failure-mode analysis, "superheater tube fails by fly ash erosion" might have a high probability and "superheater tube fails by over temperature creep rupture" might have a low probability.

At this point, the real power of risk-based methodology enters. Even at the qualitative level, using risk rather than probability or consequence cost alone enhances the discriminatory ability (and therefore the usefulness) of the analysis. This value flows from the ability that the analysis provides to look simultaneously at many components and focus on those that cause problems that have a significant effect on production.

To introduce risk, you need to assign an occurrence consequence as well as a failure probability to each component or component and failure mode combination. Recall that qualitative assessment has no way to incorporate more than one sort of consequence; you will therefore need a way to deal with non-economic consequences such as safety and environmental damage. Again, selecting "high," "medium" or "low" occurrence consequence is sufficient. You will then combine probability with consequence cost in a risk matrix. A risk matrix graphically presents risk by combining the probability and consequence estimates. Figure 5.7.1 is a risk matrix. You can rank components by risk, then select those with the highest risk for further attention, or you can rank and select the most risk-significant failure mode(s) for further study.

You've needed no tool more (or less!) complex than pencil and paper yet — but you've probably acquired significant information. Consider: although you may have learned nothing that your intuition had not already suggested, you now have organized your data in a more effective way for communication to others. Also, risk-based inspection is like any other analysis tool. The effort to properly pose a problem often solves the problem.

1.4.3 □ Quantitative Risk Assessment

All the other risk-based process modules involve quantitative risk assessment. You will replace "high,""medium" or "low" probability and consequence estimates with failure probability and consequence cost values. This refinement requires more effort and more detailed data, but the payback can be very large.

You will require quantitative system assessment if:

- The qualitative assessment did not clearly identify a course of action, or
- You have eliminated the "problem" that the qualitative assessment ranked highest and it isn't clear where to attack next, or
- You need to compare risks associated with different units or facilities, or

- You have identified several strategies that will address an issue, but a qualitative assessment does not adequately rank them, or
- You must decide which of several same-ranked projects you should pursue when you do not have the resources to pursue them all, or
- You require financial or other quantitative information to defend your recommendations(s), or
- You simply want more of the added value that risk-based methods can deliver.

Note that, as soon as the risk-based analysis moves beyond addressing a limited, well-defined problem, one or more of these results are likely.

Quantifying failure consequence cost requires some effort, but is reasonably straightforward. You may not normally use the required financial data, but you should be able to obtain all the information that you will need from sources within your organization. This is not so with failure probability information. To quantify failure probability, you will need failure probability data or engineering analysis. Ultimately, you will need failure probability vs. time data.

Two broad classes of failure probability data exist: specific data and generic data. Specific data is locally-gathered information. Data may be equipment class specific, unit specific, component specific or anything in between. Generic data is compiled information, usually from an industry or public source.

Generally speaking, the more specific the data source the better. However, specific data set use raises two issues. First, such data sets very often don't exist; where they do exist, they may not be in an immediately useful form. Second, they may be sparse, that is, they may not be statistically relevant. In fact, for low probability events, they will certainly be sparse. The fact that a given boiler has not yet had a tube rupture after five years of service assuredly does not mean that tube rupture is impossible. Equally, the fact that a boiler has had a steam drum rupture after five years' service does not mean that five years is a good steam drum life estimate.

Generic data, on the other hand, particularly industry-wide generic data, reflect a fleet-wide average of operating conditions. Some operating conditions will be better than yours and some worse. You have no way to know where in the spread your facility lies. Section 6 tells you how to obtain and use generic data to get your program started, and Section 9 tells you how to supplement generic data with data from other sources.

Section 9.4 introduces tools such as **remaining life analysis (RLA).** Section 9.4.3 tells you how to format the information that these tools can provide and use it with your system definitions and the generic or specific failure probability and consequence cost data to more precisely rank the components for the selected failure modes. You will also produce a failure probability versus time relationship. You will need this relationship in the mathematical model that you will develop for the more sophisticated risk-based analysis procedures.

Section 8 introduces fault tree and event tree analysis. When the relationship between component failure and mission failure is not clear, then you will need these tools to analyze that relationship. If you think that human factors are immediately important (e.g., an important event tree gate or fault tree basic event is a human action), then you can use the procedures in Appendix D to develop a human failure probability.

Basic quantitative risk assessment can require a variety of resources. A database can be helpful; Section 4.1 discusses useful data structures and how to construct a database. You can do extensive curve fitting and component ranking with spreadsheet software on a personal computer. The Handbook CD-ROM provides templates, or example spreadsheets, and macros that will help you to do these things. If you are analyzing complex systems, you will find fault tree and event modeling to be useful tools. The Handbook CD-ROM also provides fault tree/event tree software for your personal computer. Appendix C.1 (The Fault Tree Handbook) explains how to do fault tree/event tree analysis manually.

1.4.4 □ Multiple Component Ranking Based Upon Optimization

The next step is called multiple component ranking based upon optimization. In the initial quantitative assessment, you might do very sophisticated probability work, but you will risk-rank by using only MWH (production) loss. MWH loss is the direct unit of production loss for a fossil fuel-fired power plant. You can easily use a different output measure if you are analyzing a different kind of facility.

In this step, you will account, in much greater detail, for the economic factors and safety constraints that your organization faces. You will use decision analysis tools to evaluate inspection and replacement decisions that maximize net present value consistent with year-to-year capital availability limitations. Your model will, based on unit need, optimally balance resources among components and units and help you develop a complete productive resource management strategy.

You can develop a component replacement strategy with spreadsheet software and a statistical analysis extension on a personal computer. The Handbook CD-ROM provides templates, or example spreadsheets, and macros.

1.4.5 □ Inspection Program Development

After you have found exactly which components are most significant to NPV and in which year each important component requires attention, you will need inspection programs. Inspections in a risk-based program increase confidence in equipment reliability by determining actual conditions and thereby reducing uncertainty in the risk assessment model. For the procedures in this Handbook, an inspection strategy is a set of parameters, such as method, sample size and access method, that completely define how a component under consideration will be examined. (Recall that the timing element of the strategy came from the previous step, multiple component ranking based upon optimization.) You will develop a model, then enter inspection costs and success probabilities. Then you will use the model to find the strategy that maximizes NPV. You will also factor safety and other goals into the analysis to ensure that the optimized program will meet these goals. After inspections are performed, you will feed the results back to the model.

Inspection program development starts with a "do nothing" or base case. "Do nothing," however, really means "do nothing different." Your existing inspection program, including statutory inspections, is affecting the failure probability at some cost. If you do not consider the existing program, you might assume a "do nothing" failure probability that is too low. If you do not consider the cost of the existing program, then you will not be able to calculate an NPV, and any different program will show an inappropriate cost advantage. See also the last paragraph of Section 1.2 for a caution about undocumented work that might take place during statutory inspections.

You can develop an inspection program, evaluate it, and use its results to update a quantitative risk model with spreadsheet software and a statistical analysis extension on a personal computer. The Handbook CD-ROM provides templates, or example spreadsheets, and macros.

1.4.6 □ Sensitivity Analysis

Multiple component ranking based upon optimization and inspection program development use spreadsheet models to help you make decisions. Sensitivity analysis is a tool that will identify the variables that have the greatest effect on your decision. Section 12 tells you how to do a sensitivity analysis and suggests how you might strengthen the variables that you find are the most important.

NOTES:

NOTES:

SECTION 2

This step requires:

- *Incipient understanding of probabilistic and risk-based methods.*
- *An interest in further exploring this Handbook.*

This step provides:

- *A Handbook "verbal roadmap." (For a more visual roadmap, see also RBMAP.PDF on the Handbook CD-ROM. Note that the roadmap node colors are designed to match the colored tabs of their corresponding handbook section.) The verbal roadmap includes:*
- *A discussion of the Handbook-suggested process order and purpose.*
- *A brief description of each major Handbook step.*
- *More Handbook usage suggestions.*

The next step will, in general, be system definition.

PROGRAM DEVELOPMENT PROCESS 2

This Handbook will help you apply a selection of risk-based tools. The process in which these tools are generally applied is outlined in Figure 1.2 and is shown in greater detail in RBMAP.PDF. The process modules include:

- System definition
- System information assembly
- Qualitative risk assessment
- Quantitative risk assessment and ranking
- Fault tree/event tree modeling (when needed)
- Combining data from various sources
- Multiple component ranking based upon optimization
- Inspection program development using decision analysis
- Sensitivity analysis
- (Optional) Human factors analysis

You, an engineer who works with your facility every day, are the person who can most efficiently apply these techniques and produce an effective risk-based program. In general, it is more effective for you to learn and apply these risk-based techniques than it is for an expert risk analyst to learn the operational details of your facility. When you need to seek specialized help, having applied this Handbook will give you the "knowledge bridge" that you will need to efficiently use the help.

The basis for these techniques is discussed generically in (ASME, 1991) and conceptually for fossil fuel fired power plants in (ASME, 1994). The following sections will show you exactly how to apply these techniques and will provide examples to illustrate every step.

Section 3 shows you how to develop a system definition. You will take "your problem," which might be vaguely defined at this point, and frame it for formal investigation. You will need system block diagrams, schematics and drawings. You will also need to collect or at least identify the sources for the system and component information that you will need later, because data availability can strongly influence the resources you will need as your analysis progresses. This step will produce a formal system boundary and its associated success criteria.

Section 4 guides you through your initial system data collection and explains how you should tabulate it for further analysis. You will need system- and component-level engineering data. For each component of interest, you will also need an operating history, maintenance and repair records, and some insight into possible deterioration mechanisms and their effects. If you foresee an immediate need to use the quantitative methods, you can also collect data that will support these methods. If you find that your data cannot support the analysis that you are doing, you might need to go back and reformulate your system definition, or you will need to stop and develop the data that you need through research and analysis, local expert elicitation, or hired consultant(s). This step will provide the data you will need to begin your analysis. If your facility-specific data is not readily available or is not able to support your analysis, you may also be able to select, obtain and format generic data to get you started.

Section 5 helps you to break your system into components and then to risk-rank those components. Optionally, it helps you to look at component failures, their causes and/or effects and at component failure consequences. You will need failure probability and consequence severity estimates for each component of interest. Optionally, you will need failure probability and consequence severity estimates for all likely failure mechanisms for each component. You might find that you need to redefine your system or to collect more information. This step will help guide you to a qualitative risk ranking and risk matrix. If the qualitative assessment serves your immediate needs, you can restart the process with another issue. Otherwise, you might find the qualitative risk matrix is a useful tool for obtaining "buy-in" and support for more detailed investigation.

Section 6 introduces quantitative methods and discusses when you might need to use them. It also introduces the various analysis techniques that you might use to move from qualitative to quantitative analysis.

Section 7 tells you how to use facility-specific or generic data to estimate failure probability and to develop risk plots. You will need some basic component failure consequence information. This step will provide a "first pass" quantitative risk assessment for issues that involve no redundant trains or components. For issues that do involve redundancies, this step will provide inputs to the further analysis that you will need. In this step, you will also develop failure probability vs. time curves.

Section 8 introduces you to system functional diagrams and to fault tree and event tree modeling. These tools will help you to understand more complex systems and failure scenarios. You will also use these techniques to analyze systems that contain redundant **trains** or components. You will need information about the redundant equipment and how loss of each train or component, singly or in various combinations, will affect plant output. This step produces mathematical models that can relate component failure probabilities to system mission failure probabilities. Once you have built a fault tree/event tree model, you can multiply exercise the model to produce an estimated failure probability vs. time curve for the complex system or any of its parts.

Calculating relationships between component failure and mission failure in complex systems is just one application for fault tree and event tree analysis. These are powerful analytical tools independent of a risk-based program. If you wish to further explore fault tree analysis, see Appendix C.1 (The Fault Tree Handbook) and Appendices C.2, C.3 and C.4 (IRRAS and its manuals).

(Optional) Appendix D introduces human factors analysis. If important event tree gates or fault tree basic events are human actions, you could set their probability to 1.0 and note this in the analysis report, or you could perform a more robust human factors analysis. See Appendix D for more information.

Section 9 tells you how to develop plant-specific data, how to use expert elicitation to obtain failure probability information and how to properly combine data that has been collected from various sources. If you started your analysis with generic data, you might have or later acquire "better" data. If you find that generic failure data are critical to your analysis, then you might wish to acquire and apply data which are more directly applicable to your facility. This section will combine data from various sources in a statistically sound manner.

Section 10 shows you how to build a decision model that will perform multiple-component economic optimization and safety-constrained optimization. Up to this point, you have been building a foundation. Although the work that you have done up to now may have provided valuable insights and identified useful improvements, you have essentially considered all facilities, if not quite all components, as roughly equal contributors. Of course this is not so; to base your programs on "bottom line" impact, you will need to gather and incorporate various information about your facility, such as replacement power costs, and its planned usage, such as projected load factors. Finally, you will need some information about how your company calculates its internal costs. This step considers the risk-based models for several components and, constrained by various corporate safety and financial factors, helps you to decide when they require attention.

Section 11 shows you how to develop inspection programs. You will need information, such as detection probabilities and costs, for inspection methods that apply to the failure modes of interest. You will use strategy tables, decision trees, and influence diagrams to develop the most economical inspection program. The resulting program will satisfy the safety, budgetary and any other constraints that you selected and applied in the previous step when action is taken at the timing that was determined in the optimization. If you wish, you can modify the inspection program development step so that it applies similar constraints.

Section 12 shows you how to perform a sensitivity analysis on the models that you use in Sections 10 and 11. Such an analysis will identify which are the most critical variables in your models, and which therefore are most worth the cost of more study. This section also suggests how you might improve or supplement various kinds of information that prove to be important.

Figure 1.2 shows the major feedback loop in the process, i.e., the return to the top with newly-found knowledge. RBMAP.PDF shows many more options. Its added complexity illustrates several of the more likely feedback decision points that you will encounter. You sometimes find that you don't know as much as you thought that you did, or that the data that you are using will not support what you are doing. In either case, you can return to the definition and data gathering steps. Any time that you change your inputs or reframe your inquiry, you should, before proceeding to another module, rework any earlier modules to which the changes may apply.

Finally, you must maintain a healthy skepticism at all times. Remember that you are the best risk-based analyst for your plant because you know your plant. If you get results that "feel" wrong, try to rework the analysis from a different direction; do a sensitivity analysis to find which inputs or estimates might need refinement; get a colleague to review your model(s) with you. Risk-based methods might reveal some surprises, but you must be careful. Remember: one premise in decision analysis is that intuition is an important input.

NOTES:

SECTION 3

SYSTEM DEFINITION

This step requires:

- *One or more "problems."*
- *System and component drawings that cover the area(s) of concern. You will probably need drawings that cover the specific area of concern, one system level "above" and at least one component level "below."*

This step provides:

- *A clear, easily-communicated system boundary.*
- *An argument for a financial basis in engineering decision analysis.*

The next step will, in general, be system information assembly.

SYSTEM DEFINITION 3

You must first identify the target of your inspection program. This means that you must select the physical system that you want to investigate and determine its success criteria. You might also want to immediately consider failure modes or mechanisms. For example, should you focus your inspection program on:

- Only the portion of the facility that is presently causing trouble or should you include portions that are expected to raise issues in the near future?
- Plant availability at 100% power or plant availability at a lower power level?
- A specific failure cause across all affected components, a range of failure mechanisms for selected components, or possibly on the components themselves?

You may also want to begin considering and identifying operator safety limits, performance limits and budget constraints.

Once you have clearly defined your objective, you will be able to:

1. Create system boundaries that enclose the objective
2. Limit the analysis to activities that are needed to address the objective
3. Identify when the objective has been met
4. Rationally select component boundaries
5. Create a failure data base that will provide the necessary failure data without wasting resources by including irrelevant data.

When you have completed this step, you will have a tool that will help you to obtain agreement about the program scope that you can discuss with your supervisor and others. You may also have collected enough information to risk-rank the components of immediate interest without further analysis.

System definition in a risk-based program requires a boundary and a success measure. The boundary clearly identifies what the system is and what it is not. Think of it as a control volume like you probably used in thermodynamics. The success measure is generally referred to as the success criteria, whether it is singular or plural, simple or complex.

3.1 □ SYSTEM BOUNDARIES

Defining a system boundary is easy to do but hard to describe. Depending upon the nature of your "problem," you might want to:

- Start thinking about your facility "from the coal pile to the transmission line" and then progressively narrow your focus until you highlight all the areas about which you are concerned, or
- Start thinking about, for example, superheater tube bends and then progressively broaden the focus until you include everything in the same functional unit (i.e., boiler) that presents similar concerns.

To illustrate, consider possible system boundaries for the example "problems" in Section 1.2.

- *"How do I organize the accumulated information in operating logs and in outage, failure investigation and inspection reports? And how can I use this information after it's organized?"*

This case could be as simple or complicated as you wish. You could define your system to include all the components for which you have data, or you could narrowly define the individual systems that encompass each data "cluster," one at a time, possibly tying the clusters together later.

- *"I've been assigned to reduce the forced outage rate for our most efficient unit. What component, system or procedure do I study first?"*

Define the unit as the system. Define systems and their components consistently with the data that you will use, whether it is generic or unit-specific. You will need to draw system boundaries around components consistently with the data that you have, or you will need to commit resources to develop additional and/or more detailed data.

- *"Management has cut the maintenance budget. Can I safely achieve the required unit forced outage rate limit with the available resources? If so, how? And if not, how do I justify more resources?"*

Same as the previous, but now you involve corporate value-measure considerations. You might be able to analyze only your unit, but you will also need to consider corporate-level effects, such as dispatch plans and replacement power costs.

- *"My company has four units over thirty years old. Major components have dwindling remaining lives. In what order and when do I inspect or replace them?"*

Same as previous, except that your system now should include the four units and the analysis will involve corporate-level information.

- *"I learned about a better nondestructive examination (NDE) method for boiler tubes. Should I use it?"*

This case could be as simple or complex as you wish. You could assume a test case, for example, waterwall tubes in your most efficient unit, and base the decision upon those results. However, assuming that the new method is more expensive, it might be justifiable for your most efficient unit but not for less efficient units.

The first two examples are excellent risk-based methods applications because they let you start small; get some results with qualitative assessment first, and then grow into more sophisticated methods. The other three examples have strong economic elements; they push you directly into quantitative methods. The last example, although it has an economic element, still provides an opportunity for a preliminary analysis and might stress inspection program development.

Note that safety is not generally an issue in system boundary selection. Without risk-based methods and with risk-based methods up through the qualitative level, safety lies in the deterministic arena, covered by code construction and jurisdictional rules. Qualitative risk-based methods may improve safety indirectly by focusing resources on the most risk-significant components. Quantitative risk-based methods, at least as they are presented here, are the only effective way to directly ensure that **safety limits** are met. The optimizers work on value, for reasons that are explained elsewhere, but they all reject solutions that do not comply with the selected safety limit.

Here is the procedure for system boundary establishment:

1. Identify and bound main systems. Include subsystems and parallel trains.
2. Mark the boundaries on a simplified process diagram or unit print. You must identify significant support systems such as electrical, air, control, water, and lubrication, however, these support systems and their boundaries might not be easily defined. Remember that support systems often present common cause failure mechanisms, e.g., single-component support system failures can cause multiple failures in the supported systems.
3. Rather than marking boundaries on complex drawings, consider using descriptive sentences to mark boundaries, for example, groups of steam-handling components or groups of water-handling components. Words may also be more effective than pictures in areas such as equipment interfaces, to show, for example, whether pumps include their motors and/or motor control center, or whether valves include their air operators. In fact, after you become familiar with risk-based methods, you might want to develop simplified, color-coded drawings that specifically meet your program's needs.
4. Electrical systems provide some support functions that are plant-wide and others that are mechanical-system-specific. Draw electrical systems boundaries to accommodate your interests or goals. For example, for the power to a mechanical component, you could place the boundary at the breaker in the motor control center, at the connection to the motor or between the motor and the mechanical portion of the component.
5. Consider air systems the same way, or consider them by pressure

rating or by function, such as soot blowing, control air, or service air, and select their boundaries accordingly.

6. Decide what part of the instrument/control systems and how much human interface you wish to include with each system.
7. When you define system and sub-system boundaries, consider both the program objectives and failure data availability. Avoid either excess modeling or excess failure data collection/development. For example:
   - If the available data does not differentiate between the mechanical components, their operating devices (air or electric actuator or driver) and the interconnecting components, making the boundary pass between these items would increase the cost of developing the failure data.
   - If the failure database doesn't differentiate between certain boiler tube or pipe segments, it would be difficult to support boundaries between such segments.

In either case, however, solving your problem might, after initial study, prove to require the more difficult approach.

You should also consider alternate and future uses for your analysis and secondary concerns that it could help to address. You might be able to configure the analysis to accommodate additional needs with few or no additional resources. The inverse could also be true; focusing the analysis by limiting its boundaries could substantially reduce the amount of resources needed. Consider:

1. Does your plant have concerns with control systems, such as aging control systems and their affects on plant/system operation or plant availability? For example, including high level control input might be easy, however, incorporating detailed control interfaces might produce a complex model that requires failure data that is not available.
2. Can you make this model help define and resolve obvious plant maintenance issues on specific components? For example, you might include fuel-handling systems, which typically have high failure rates. Because fuel-handling systems typically have high failure

rates, failure data is usually available. (In the Niagara Mohawk experience that is mentioned in the Executive Summary, improving immediate problems with boiler tubes immediately "laid bare" serious fuel system problems. These problems had been hidden because the frequent shutdowns caused by tube failures allowed fuel system repairs to stay out of sight.)

3. Could you make this model support the testing/maintenance program in addition to the inspection program? For example, could it identify needed tests and monitor their effectiveness?
4. Could this model help your engineering group make design decisions by quantifying anticipated system/plant availability improvements? Most design changes claim, in a qualitative manner, to provide such improvements. A (finance-based) quantitative analysis can provide you with a forceful argument during the approval cycle and prioritization of modification packages.

3.2 □ SYSTEM SUCCESS CRITERIA

What constitutes success depends upon what constitutes failure. There might not, however, be a direct relationship between component failure and system failure. Also, whether a component failure causes system failure could depend upon the operating mode at the time of failure.

Success criteria are generally operation-based, failure mode-based or component-based, however, any success criteria should include a "unit mission" component, such as maintaining 100% power, maintaining 40% power, or avoiding forced outages. When you identify the unit mission, you identify the specific failures or failure combinations with which you will be concerned. For example, with a three-train feedwater system, which mission success criteria you select might define whether your analysis is about one feedwater pump or pipe failure or a combination of two or three feedwater pump or pipe failures.

You could ignore the unit mission if you are working a tightly-defined, qualitative assessment. For example, you could define success as "no boiler tube failures" or "no creep-rupture failures." Unless you "tie in" to unit mission, however, you have no way to calculate the business interruption (replacement power) part of the failure consequence. Business interruption is almost always very important in a quantitative analysis.

If you identify both component-level and unit mission-level success criteria, you might find the relationship between the two is complex. If so, you might need fault tree/event tree analysis to study the relationship (See Section 8).

3.3 □ EXAMPLE

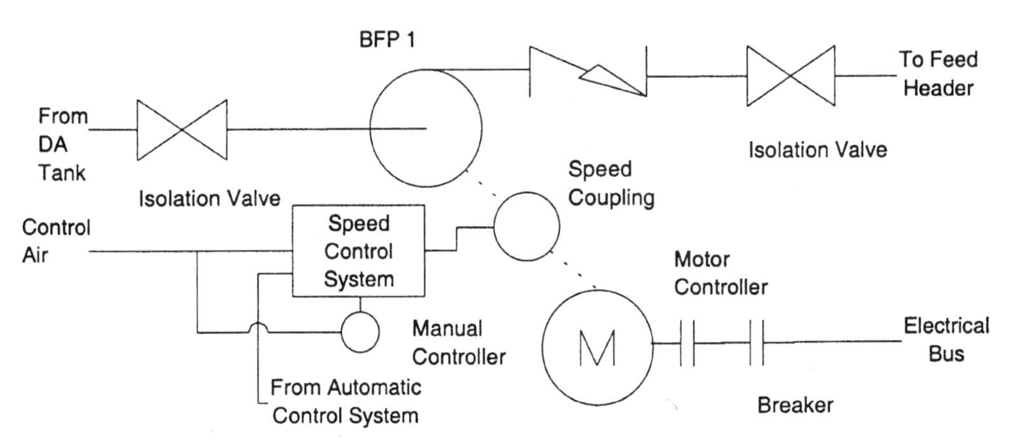

Figure 3.3.1 Main Feed Pump Train

Figure 3.3.1 shows one of three boiler feed pump (BFP) trains. One train is required for 0–55% power and two trains are required for 56–100% power. Therefore, the success criteria are 1/3 running/ready for any power production/availability and 2/3 running/ready for 100% power production/availability.

NOTES:

SECTION 4

To perform this procedure, you will need:

- *A defined system, fully identified, with logical and consistent identifiers attached to each train (if applicable) and component.*
- *Operating and failure information for the system and its components.*
- *(Optional) Source(s) of data that correlate component failure rate, component operating history and forced outage hours.*

This procedure will produce:

- *Tabulated plant-specific failure and operating history data.*
- *(Optional) Plant-specific component failure data that is formatted for quantitative analysis.*
- *(Optional) Generic data that is formatted for quantitative analysis.*

The next step will, in general, be qualitative risk assessment. You could also proceed directly to failure data analysis.

SYSTEM INFORMATION ASSEMBLY 4

Once you have identified a system that contains "your problem," you will need information about the system, its trains (if applicable) and components. The amount of detail that the information must contain is a function of the depth of analysis that you are contemplating. You might be able to qualitatively rank the components with your own knowledge. However, qualitative work is very unlikely to support all your goals for very long.

The previous step directed you to locate your system and component boundaries with an eye on the available data. Now you need to organize that data so that you will later be able to extract the information that you will need. Qualitative work might need very little formal support for your own knowledge. Quantitative work will at least need enough information to develop failure probabilities and consequence costs for all the components of interest. Full maintenance optimization will require failure probability vs. time curves and could also need detailed engineering information about the important components.

This section describes typical data structures and explores various data collection issues. You can proceed from here to qualitative risk assessment (Section 5) or you can skip directly to failure data analysis (Section 7).

4.1 □ DATA COLLECTION AND TABULATION

As you find, assemble, correlate, and analyze plant, system and component design, operation and failure information, you will be working with engineers, operators and maintenance personnel. This interaction will help these groups to understand your analysis as it unfolds and to accept its results. The time you spend talking to as many individuals in these groups as possible, gathering information from them and explaining the risk-based analysis processes to them, will be generously repaid by a more complete and effective analysis and by easier acceptance when you present the results.

This psychological aspect is important because it involves ongoing support and eventual buy-in by those upon whom the success of your (and their) program will depend. Risk-based analysis, particularly at the quantitative level, can produce some surprising results. If those who need to support those results can see "their" data going in, then they are more likely to accept what comes out, once you have explained the process.

The same people should develop or help to develop all the system models (such as reliability block diagrams and fault/event trees). These models should, as much as possible, be developed in parallel. This helps to ensure that the boundaries, component failure data, system functions, and nomenclature that you develop during each process are consistent. You can make changes immediately rather than spending time redoing a completed model or database.

Before you begin to collect data, you should examine the way that your plant identifies equipment. Not all identification systems will support your program. For example, physical locations relative to columns and elevations are difficult to use. Further, you can expect equivalent components in parallel or similar trains to have significantly different failure rates. Therefore, you might need to track more equipment than the existing system separately and explicitly identifies.

You might need to develop your own scheme that uniquely identifies systems, trains and individual components in a logical, purpose-oriented manner that is consistent with a database or spread sheet and computer sorting. One common system uses three-part identifiers such as "BF-n-xxxx," where "BF" is the system (boiler feed, in this case), "n" is the train and "xxxx" is a component. There are many reasons other than risk-based maintenance application to identify equipment in this way. Your program could be the incentive that gets management to support a suitable identification system for a facility that lacks one.

If you are not already on intimate terms with the available data, collecting the data and building the database might be a helpful interactive process. Keep these cardinal rules in mind:

1. Section 3.1 explained that the nature of the available data strongly influences the analysis that it can support. In the long run, the quality of the currently-available data cannot control the analysis, however, decisions to expand the analysis beyond what the existing data can support, e.g., by extensive/expensive research and/or by changing the data gathering systems, are not to be taken lightly. You might want to note what you think you need to know, start working with what you have, and let the sensitivity analysis that you will perform later help you to decide which data gaps are the most important ones to fill. You can even use your risk-based model to calculate the expected value for the proposed data so that you can do a cost-benefit analysis.
2. You must control the allowable input for each field, even those that appear to be "free-form." "Air tank" and "air receiver" and even "air tk" and "air tk." are all different to a computer, as are "rupture" and "burst." If such distinctions aren't important, then don't capture them.
3. Avoid like the plague entries such as "NOC" (not otherwise classified) or "unknown." These items have a nasty habit of being, for example, dominant failure causes.
4. Be as concise as possible, but avoid having to make multiple passes through volumes of paper to collect information that you could have picked up on the previous pass.

You should explore all local failure data sources. You will find some failure data in operating and maintenance logs and reports. Failures that caused plant outages or plant de-ratings may have been reported to North American Electric Reliability Council - Generation Availability Data System (NERC-GADS); you might obtain this data from the person who prepares the NERC-GADS reports. Other failure information may exist uniquely on some individual's bookcase or in some individual's memory.

TABLE 4.1.1 TYPICAL DATABASE STRUCTURE

Field	Contents	Entry	Meaning
{SYSTEM}	System name	FUEL	–
		FW	Main feed water
{F_DATE}	Failure date	07/05/85	–
{REC_NO}	Record number	92978	–
{COMP_ID}	Component ID	04-0003-MILL	Train 4, s/n 0003, coal mill
		02-0005-BFP	Train 2, s/n 0005, feed pump
{COMP}	Component	HGVALVE	Manual gate valve
	type	PTVALVE	Air op. globe/throt. valve
{FAIL_SEV}	Failure severity	C	Catastrophic
		D	Degraded
		I	Incipient
{OP_STAT}	Operating	T	Failed while running
	status	D	Failed to start
{CAUSE}	Cause	25	Table lookup
{R_TIME}	Repair time	42	Hours to restore
		Optional Fields	
{AGE}	Age at failure	3574	Operating hours
{STARTS}	Number of starts at failure	146	–
{MWH_LOSS}	Production loss	12600	–

Table 4.1.1 shows information that might appear in a database. Note how the fields might be used:

1. {SYSTEM} is a good failure data-sorting field that will support model development and will help ensure that the system model is relevant and complete.
2. {F_DATE} (Failure date) can help you identify and delete duplicate entries for the same failure. This field also allows you to combine information on the same failure that came from different sources. It also positions the failure relative to past events, such as design modifications, component replacements or operating mode changes.
3. {REC_NO} (Record Number) can provide the traceability to source documents that will demonstrate that the database is credible. This field also permits you to research a specific component failure at a later date.
4. {COMP_ID} (Component Identifier) connects the component to a specific "part" in a specific unit or system train.
5. {COMP} (Component type) permits you to sort when you are looking for failures common to a component type.

6. {FAIL_SEV} (Failure severity) helps you identify the effects past failures have had on the systems. You can obtain some insight into the effects on system or plant operation that were caused by changing operating, inspection or testing programs.
7. {OP_STAT} (Operating status of the component at time of failure) (e.g., Fails to Start or Fails when Running) helps when you are developing inspection or testing strategies. It also helps you to develop failure models that reflect the actual situations in which failures are occurring.
8. {CAUSE} (Failure cause) helps you to understand what is actually happening in the system and to the specific components. In most cases, a root cause will not be identified. However, report what the reporting document states, even if it is "broken," because the number of failures that you will be entering will probably make re-analyses of all the failures impractical. You can reanalyze specific failures of interest later.
9. {R_TIME} (Time (required) to repair and return to service) provides you with down time. However, always watch for abnormalities such as repairs that were intentionally not performed for months or years, or inappropriate temporary repairs.

After you have gathered the failure data, estimate the operating time and the number of start-ups from operator logs. Tables 4.1.2 and 4.1.3 are typical operating times and numbers of start ups for feedwater pumps in a fossil-fueled power plant. In a single year and between years, these numbers can obviously vary greatly for identical components. For example:

- The hours of operation range from 7 hr/yr./pump to 8,040 hr/yr./pump in 1995.
- The total number of starts ranges from 37 starts/pump in 1995 to 273 starts/pump in 1992.

Thus, an increased number of failures for a specific feed water pump may actually represent an improvement if the hours of operation and number of starts increased greatly.

TABLE 4.1.2 BOILER FEED PUMP OPERATING HOURS (TYPICAL)

	Operating Hours			
Year	BFP-1	BFP-2	BFP-3	Plant Hours
1992	0007	5996	5586	7278
1993	4290	7914	4470	8454
1994	2475	1146	3899	3830
1995	7861	0007	8040	8001
1996	4004	3719	7579	7673
Total	18637	18782	29574	35236
Average	3727	3756	5915	7047

TABLE 4.1.3 BOILER FEED PUMP STARTS (TYPICAL)

	Operating Hours			
Year	BFP-1	BFP-2	BFP-3	Plant Hours
1992	1	104	168	7278
1993	8	19	49	8454
1994	13	8	24	3830
1995	16	4	17	8001
1996	6	47	14	7673
Total	44	182	272	35236
Average	9	36	54	7047

4.2 □ GENERIC DATA

When there is no local, facility-specific, component-specific engineering analysis or local expertise upon which to draw, the next best resource is the experiences of other operators of similar equipment. Databases built on these experiences allow you to logically project the possible future performance of your equipment.

Unfortunately, generic data reflects the worst experience of others; that is, the data is biased toward the worst-performing units in the fleet. The first units to fail are the members of the population that have the shortest life performance, whether that performance was caused by "infant mortality" or by substandard design, environment or operating practices or by some other adverse influence. Your equipment might not be subject to these influences; for a variety of reasons it may be on the long side of the failure probability distribution. This means that the analyses you perform based upon such data could be overly conservative and therefore unnecessarily costly to your company. However, if generic data is the only available information, then it is the best you can do as a starting point. You can refine the data later, if necessary, when you have accumulated more data specific to your plant, or you can use expert opinion and/or engineering analysis results to refine the data.

There are a number of sources that can provide data of the type that you need. Some are:

- North American Electric Reliability Council - Generation Availability Data System (NERC-GADS)
- Insurance company databases
- Manufacturer's databases
- Specialized databases

These databases each have positives and negatives.

The NERC-GADS is the most easily accessible for utilities, however, it does not go into detail on forced outage causes for specific components. It only identifies the component that caused each forced outage. This database does, however, contain the component calendar age at failure.

Insurance company databases are not necessarily accessible to anyone outside the company that owns them, and they are biased, particularly in that they only reflect events that lead to insurance claims. This means that they may only include events that caused losses above the insured's deductible. Further, most policies on utilities do not include business interruption. Therefore, lost generation costs will not be included in the data. Many small but important losses, such as boiler tube failures, are unlikely to show up. An advantage is that insurance company databases may contain more detailed component failure causes than the NERC-GADS provides.

Manufacturer's databases and specialized databases also have strengths and weaknesses. If you operate combustion turbines, you should explore the ORAP database, especially if your company is an

ORAP member.[4] If any other databases are available to you, you will have to analyze them on a case-by-case basis. If you have access to the data and can extract it in the form that is described in the following discussion, you can use the data directly in your analysis or you can use the procedures in Section 9 to combine the data with data from other sources.

The basic details that a failure history database needs to include are:

- Component identifier
- Operation age of the component at failure
- Number of component starts at failure
- Magnitude of lost production caused by failure
- Time duration of lost production caused by failure

You need these details to help you to determine the relationship between the component operating history and its failure history. When you know about a component's past performance, you can logically project its future performance. However, remember: when you make a projection that is based upon past operating history, you are implicitly assuming that the future operation will be similar.

The examples in this Handbook will use the NERC-GADS database, primarily because it is accessible and because it records all forced outage events.

4.3 □ NERC-GADS DATA

The NERC-GADS database is a compilation of forced outage event data for power plants all across North America. Utilities have been contributing to this database since the mid 1970s. It has existed in electronic form since 1982. The forced outage events are identified by cause codes that call out the responsible component for each forced outage. This fact makes the database a very useful generic failure data source. Two other databases operate parallel to this outage event database. One is a unit

[4]For information about ORAP, contact Strategic Power Systems, Inc., 11301 Carmel Commons Boulevard, Piedmont Building — Koger Center, Suite 308, Charlotte, NC 28226-3976, Telephone 704-544-5501, Facsimile 704-544-5505.

performance database that describes the unit operation information. The other is a design database that records unit design, manufacturer and commission date.

Two ways of using the NERC-GADS database have been investigated:

- Direct extraction from the raw databases and
- Indirect extraction using a NERC-GADS database interrogation program called pc-GAR.

Extracting the raw GADS data is difficult. First, you must pre-format the raw data with a specially written FORTRAN preprocessing program on a large workstation. Then you must further process the data with PC database and spreadsheet programs (e.g., Microsoft Access and Excel) using sophisticated queries and macros.

Using the pc-GAR program, which is available from NERC on CD-ROM, to process the data is a better approach to this data acquisition effort. This is especially true if resources are constrained and/or the person performing the analysis is not fluent in FORTRAN and in the necessary PC software. When you use this technique, you will make some implicit assumptions, which will be described later, about the data. The accuracy that you will lose by making the assumptions is more than paid for by the resources you will save by using the data. Remember that generic data are a starting point and not a final solution in any case.

To obtain the pc-GAR program or inquire about the raw NERC-GADS data, contact the manager of the GADS database at (609) 452-8060, or write to:

Manager-GADS Services
North American Electric Reliability Council
Princeton Forrestal Village
116-390 Village Boulevard
Princeton, NJ 08540-5731

There is one other issue with raw NERC-GADS data and pc-GAR data: the cumulative operating hours for the unit at the time of failure is not available. To address this, you can extract the data on a calendar year basis, called a "Study Period", for all the years of the NERC-GADS database. The most direct approach is to extract data for units with the same installation (commercialization) year as your unit. This will select units from the NERC-GADS database that most closely approximate the age of your unit. You will extract data for each cause code of concern. The multiple runs can be time consuming, however, multiple cause codes can be extracted during each run.

To obtain the forced outage numbers, hours and lost production (MW hours) for each cause code for each calendar year of the database (study period) for units of a specific installation year, proceed as follows:

1. Start pc-GAR
2. Answer [Clear Previous Session Reports?] by clicking [Yes].
3. Click on [Criteria] on the menu.
4. Select [Study Period].
5. Type in the 1st month and year of the period and last (12th) month and year of the period. If your unit was commissioned before 1982, then start with 1982. (This period will be incremented to get the number of forced outages by year of unit age.) Then click [OK].
6. Click on [Specific Units] on the menu.
7. Select [Other Units].
8. Select [All Units, Including Own].
9. The program will process the data and output the total number of units for that study period.
10. Click [OK].
11. Click [Finished] to return to the main menu.
12. Select [Design Parameters].
13. Highlight [Fossil Steam] and click on [Select].
14. Highlight [General Information] and click on [Select].
15. Highlight [Commercial Date] and click on [Select].
16. Type in beginning and end dates for the year of the commercial date year in which you are interested.
17. Click on [OK].
18. Click on [Continue] in the [General Information] box.
19. Click on [Continue] in the next two boxes.
20. The program will process data and output the number of units that were

selected. Note the number of units. You will need the number of units commercialized in the year of interest later in the analysis

21. Click on [OK].
22. The program will further process data and return you to the main menu.
23. Click on [Operating Data].
24. Click on [Performance].
25. Highlight and select [Forced Outage Hours].
26. Enter a lower limit of 0.0 and an upper limit of a large number like 100,000.
27. Click [OK].
28. Highlight and select [# of FOH Occurrences].
29. Enter a lower limit of 0.0 and an upper limit of a large number like 10,000.
30. Click [OK].
31. Click [Continue].
32. After returning to the menu, click on [Sum Data].
33. The program will further process data and output the number of units selected. Click [OK].
34. When you are returned to menu, click on [Finished], then click on [Cause Codes].
35. Highlight the cause code range that you desire and click on [Select]. See Appendix E for a listing of the cause code descriptions.
36. Under [Select All] check [No].
37. Click [OK].
38. Highlight the cause code(s) desired and click [Select/Unselect].
39. Click [OK].
40. Repeat steps 35 through 39 until you have selected all the cause codes that you wish to process.
41. Click [Continue] and the program will process data.
42. When the processing is complete, the program will output the number of cause codes that were processed. Click [OK].
43. When you are returned to main menu, click on [Reports/Tables].
44. Click on [Reports].
45. Highlight [Individual Cause Codes].
46. Click [Select/Unselect].
47. Click [OK].
48. When you are returned to the menu, click on [Execute].

49. Click on [Prepare Output Data].
50. After processing is complete, click on [Produce Reports].
51. You can either view the information on a screen or print the [Design/Performance Criteria Report]. When you are finished, click on [Exit].
52. When the program returns the number of events selected, click on [OK].
53. Next check [Yes] on [Detail Report] and [No] on [Summary Report] and [Export CSV].
54. Then click [OK].
55. You can either view the information on a screen or print the [Individual Cause Code Detail Report]. When you are finished, click on [Exit].
56. When you are returned to the main menu, click on [Exit].
57. Then click on [Exit to DOS].
58. To obtain data for the next study period or year for units of the commercial year of interest, restart pc-GAR at step one.

4.4 □ INTERNAL COMPANY DATA

If your company is a NERC-GADS member, it may have retained component/unit-specific data that was the basis for its NERC-GADS reports. To locate this data, find out who sends your company's data to NERC-GADS. You will probably find that the data is already in the NERC-GADS format. Such data are better than the NERC-GADS data because they will be specific to your plant or company. Also (avoiding the problem with NERC-GADS data and pc-GAR), you will have access, from other company resources, to the component age in operating hours for each failure data point. If your data are too sparse to stand alone, you could use the NERC-GADS data and use the procedures in Section 9 to combine it with your company data. Once you have the data in the correct format, you can apply the techniques that follow.

Even if your company is not in the NERC-GADS, it may have component/unit-specific outage data. If so, you will need to search for its location. You will probably need to reformat the data. Such data are better than the NERC-GADS data because they will be specific to your plant or company and will include the component age in operating hours for each failure data point. If your data are too sparse to stand alone, you could use the techniques in Section 9 to combine your company data with the NERC-GADS data. Once you have the data in the correct format, you can apply the techniques that follow.

4.5 □ (OPTIONAL) DATA FORMAT FOR FURTHER ANALYSIS

Strictly speaking, this step is optional only if you are working your first pass through the risk-based process. Sooner or later, you will want facility-specific failure data to support an optimal component life management program.

There are two uses for such data in quantitative risk-based methods, ranking and analysis. To use your data in a quantitative risk ranking, you will need a spreadsheet with the following columns:

a) {Component identifier} This is the type of component. How specifically you identify components will depend upon your data and your analysis needs. As a rule of thumb, if you have fewer than ten failures for a component type, you should probably broaden your identification scheme or plan to combine your facility specific data with data from other sources.
b) {Age} The operation age of the component at failure in years. You will need to use the operating hours that you calculated or estimated from the operating logs and records to calculate this. You will need one record (e.g., one row) for each operating year for which you have recorded one or more failures.
c) {Forced outages} The number of failures for the component in this operating year.
d) {Production loss} The production demand (MW or other suitable unit) during the operating year. If you don't know the demand, use the unit maximum dependable capacity.
e) {Lost time} The total time duration of lost production (hr) caused by the failures of the component this operating year.

Section 7.1 tells you how to use this data for risk ranking.

If you are going to use your data to develop plant-specific failure rate vs. time curves that you will need for quantitative analysis, you will need a spreadsheet that contains items a), b) and c) from the previous list:

Section 7.3 tells you how to generate the curve(s). You can use this data directly or you can follow the instructions in Section 9 to combine it with other kinds of data.

NOTES:

NOTES:

SECTION 5

To perform this procedure, you will need:

- *The identified system of interest, broken down into trains (if applicable) and components. See Section 3.*
- *Component failure rate and consequence severity information.*

This procedure will produce:

- *A qualitative risk assessment.*

The next step will, in general, be quantitative risk assessment for the selected components. Qualitative risk assessment might require a return to system information assembly. Or, if the qualitative assessment solves "your problem," than you get to start over with another "problem."

QUALITATIVE RISK ASSESSMENT 5

A qualitative risk assessment ranks systems and components relative to each other. When you perform a qualitative risk assessment, you assign relative failure probabilities and consequence severities in broad groups, such as "high," "medium" and "low." Although you can use any number of groups, you will probably not be able to assign, with sufficient confidence, more than five failure probability and consequence severity groups.

Although a qualitative risk assessment provides useful information for a limited population, such an assessment generally does not easily provide sufficient discrimination over a large population and generally cannot be used to reliably compare risks associated with different systems or facilities. Different systems and facilities generally have different consequence bases and could have different probability bases. Within these limitations, qualitative risk assessment is a useful component failure-ranking tool that effectively communicates its results.

Qualitative risk assessment features speed and relative simplicity compared to quantitative methods. If you have a broadly based problem, once you have defined the system (Section 3), qualitative risk assessment provides an easily communicated, supportable way to ensure that you have "covered all the bases" in your preliminary work. You may focus your inspection resources on the basis of the qualitative assessment, or you may use the assessment to decide where a more detailed analysis might be useful.

In this Handbook, we use qualitative risk assessment primarily as a screening tool, however, it has been further developed and applied. (AIChE 1992) provides a number of examples of qualitative risk matrices and references documents that further discuss risk matrices.

5.1 □ QUALITATIVE ANALYSIS METHODOLOGY

The first step in a qualitative risk assessment is choosing an analysis method. If a formal hazard analysis has been performed at your facility, such as a failure modes and effects analysis (FMEA), hazard and operability study (HAZOPS), or fault tree analysis (FTA), you might be able to use its results in your assessment. However, for the purposes for which this Handbook recommends qualitative risk assessment, less formal tools are satisfactory. A semi-formal "what-if" analysis technique is generally sufficient. The most plausible information sources are your own experience and the experiences of operators and other plant personnel who have some depth of knowledge and experience with the equipment.

If you have formal analysis results, use them as applicable in the following steps. If you don't, then follow the suggestions that are listed at each step.

5.2 □ TABULATING SYSTEMS AND COMPONENTS

1. List the systems, trains (if applicable) and components and the success criteria that you identified in section 3.
2. Working through the system, verify that all the trains (if applicable) and components that could affect whether the success criteria are met have been listed.
3. Work through the list once more, and eliminate any items found that can not affect whether the success criteria are met.
4. Using the system success criteria as a guide, determine the success criteria for each train (if applicable) and component. For example, identify any redundancy, e.g., multiple trains or installed spares and, if applicable, determine the fractional contribution to success that is required of each.

5.3 □ IDENTIFYING FAILURE CAUSES (OPTIONAL)

You now have another decision to make about the analysis depth that you desire. You might want to simply base your analysis on "go-no go" or "failure-no failure" of the previously-tabulated components. However, the qualitative analysis technique offers somewhat more depth by allowing you to consider failure modes, mechanisms and causes.

A failure mode is the effect by which a failure is observed to have occurred, for example, rupture of a boiler tube. A failure mechanism is the degradation process that leads to failure, for example, local overheating leading to stress rupture of a boiler tube. A failure cause is a condition or event that leads to failure, for example, a blockage that causes local overheating leading to stress rupture of a boiler tube.

Therefore, if you wish, for each item in your tabulated list, identify possible failure modes, mechanisms and causes.

5.4 □ ESTIMATE FAILURE PROBABILITIES

This is a qualitative assessment; therefore, you will rank the failure probability of each item, or of each failure mode, mechanism or cause, *relative to the other items in the list*. The last point is important because it is both the strength and weakness of the qualitative method. It is a strength because it reduces your scope and therefore makes your ranking possible. It is a weakness because all the items in your list might be probable or improbable in an absolute sense. In fact, you may prefer to think of this step as "likelihood ranking" rather than as "estimating probabilities."

Table 5.4.1 provides word descriptions and estimated probabilities on a global basis. These probability categories might be useful for a corporate or facility-wide analysis, however, you are more likely to be operating over a small part of the range covered by this chart. If so, do your probability ranking based upon the number of categories, generally three to five, with which you feel comfortable. For example, "very low," "low," "medium," "high" and "very high" are about as many categories as most qualitative risk assessments can handle.

Do not neglect to carry the distinction between "failure during operation" and failure-on-demand or "fails to start." Operational failure prevention generally involves inspection, on-line monitoring or redundant equipment, and on-demand failure prevention generally involves testing.

Assign a probability or likelihood value to each item in your tabulated list.

TABLE 5.4.1 WORD DEFINITIONS FOR ESTIMATING THE FAILURE PROBABILITY FOR COMPONENTS WITH LIFETIMES IN TENS OF YEARS (ADAPTED FROM ASME, 1991)

Possible Qualitative Rank	Definition	Failure Probability (per year)
Very High	An event which may be expected to occur more than once during the component lifetime	10^{-1}
High	An event which may be expected to occur once during the component lifetime	10^{-2}
Medium	An event which is not expected to occur during the component lifetime; however, when integrated over all system components, has the credibility of happening once.	10^{-4}
Low	An event of such low probability that an event in this category is rarely expected to occur.	10^{-6}
Very Low	An event of such extremely low probability that an event in this category is considered to be incredible.	10^{-8}

5.5 □ ESTIMATE CONSEQUENCES

You need to estimate consequences the same way that you estimated probabilities. Table 5.5.1 gives word descriptions and estimated dollar costs for several categories on a global basis. You will more likely use analysis-specific categories like very low, low, etc. However, there is a complication: safety. In quantitative risk assessment, the Handbook procedures and examples recommend an approach to safety that involves a limit on failure probability. That approach won't work in qualitative risk

TABLE 5.5.1 EXAMPLE VALUES FOR ECONOMIC LOSS CONSEQUENCES (ADAPTED FROM ASME 1991)

Possible Qualitative Rank	Definition	Estimated Consequence Cost ($)
Very High	Failure causes significant potential off-site and facility or system failure costs and potential for significant litigation.	10^9
High	Failure causes indefinite shutdown, significant facility or system failure costs, and potential for litigation.	10^8
Medium	Failure causes extended unscheduled loss of facility or system and significant component failure costs.	10^7
Low	Repair can be deferred until a scheduled shutdown, some component failure costs will occur.	10^5
Very Low	Insignificant effect on operation.	10^4

assessment because qualitative risk assessment doesn't deal with absolute probabilities.

There are several ways that you can deal with this issue. You could do two qualitative assessments, one with safety consequences and one with economic consequences. Or, you could do an economic-based assessment and flag for special consideration any items on your list that have safety-related consequences. Or, particularly if you operate in one of the European countries that has a statutory safety-financial equivalent, you could factor that into your consequence table. In any case, qualitative analysis cannot deal with safety "automatically." You must deal with it in accordance with your corporate policy.

Assign a consequence value to each item on your itemized list.

5.6 □ DRAWING THE QUALITATIVE RISK MATRIX

You will enter the items on your list on an n×n grid, where "n" is the number of probability and consequence categories or bands that you are using. Arrange the grid or matrix so that probability increases in the vertical direction upward and consequence increases in the horizontal direction to the right. In other words, an item on your list that has the highest probability and the highest consequence would be placed in the upper right-hand block.

Because this matrix combines failure probability and consequence severity, it is a risk (or risk-ranking) matrix. Increasing risk follows a diagonal line upward to the right. The example that follows shows a completed qualitative risk matrix.

Although a qualitative risk matrix isn't likely to provide very many surprises, it can be very useful as a conflict-resolving device and as a very lucid presentation of its conclusions. The matrix clearly shows your highest priority items at the upper right hand portion. You might want to further examine:

- The items on the right side of the matrix, for unacceptable consequences regardless of probability and
- The items in the upper left part of the matrix, for unacceptable nuisances regardless of consequences.

5.7 □ EXAMPLE

Note: This example is entirely hypothetical. Although it is plausible, it is shamelessly "rigged" to make the point that qualitative assessment can be an efficient screening tool.

Tubular component failures often contribute significantly to the facility forced outage rate. We will develop a qualitative risk matrix that could help us to focus and prioritize a more in-depth analysis.

Step 1: Tabulate systems and components.

In this case, there is one system, and it is the complete generating unit. There are many tubular components, so we will exclude any that cannot immediately cause a forced outage or derate. We will consider the following list:

- Boiler: Reheater tubes, superheater tubes, waterwall tubes, economizer tubes.
- Feedwater heater
- Condenser
- Turbine/generator: Lubricating oil (LO) cooler, hydrogen cooler

Step 2: Identify failure causes.

This is a preliminary screening and we choose not to concern ourselves with failure modes or causes. We will use a "go-no go" criterion. Rupture, plugging, excess leakage, etc., all are the same as far as we are concerned.

Step 3: Estimate failure probabilities.

We asked each of four experienced operators to estimate how often tubular component failures caused forced outages. The results are summarized as follows:

- More forced outages were caused by reheater tubes than anything else in the facility.
- Feedwater heater and superheater tubes caused forced outages "regularly."
- Condenser tubes and waterwall tubes failed "sometimes."
- Only two operators could remember an economizer tube failure.

TABLE 5.7.1 WATERWALL TUBE SCENARIO PROBABILITY AND CONSEQUENCE ESTIMATES

Component	Probability	Consequence
Reheater tubes	Very High	High
Superheater tubes	High	Medium
Waterwall tubes	Medium	Medium
Economizer tubes	Low	Medium
Feedwater heater tubes	High	Very Low
Condenser tubes	Medium	Low
Lubricating oil cooler tubes	Very Low	High
Hydrogen cooler tubes	Very Low	Very High

- No one could remember a lubricating oil or hydrogen cooler tube failure.

Note that, for a qualitative assessment, we did not need to be concerned about numbers. Our answers clearly fall into five categories, which are identified in Table 5.7.1. If they had not been so easily ranked, we could have gone back to our operators and asked more focused questions, e.g., "Do waterwall tubes or economizer tubes fail more often?" We could also have started an operating log and record search, however, doing so would defeat our use of the qualitative assessment as a "quick and dirty" screening tool.

Step 4: Estimate failure consequences.

We applied our knowledge of the facility as follows:

- Feedwater heater tube leaks reduce efficiency and at worst require heater isolation and unit derate.
- Condenser tube failures seldom cause forced outages but they do cause water chemistry problems until they can be plugged during a scheduled outage.
- Superheater tube, waterwall tube and economizer tube failures all cause forced outages.
- Reheater tube failures cause significantly more forced outages than the other boiler tubes.
- Lubricating oil cooler tube leaks can potentially cause turbine bearing damage.

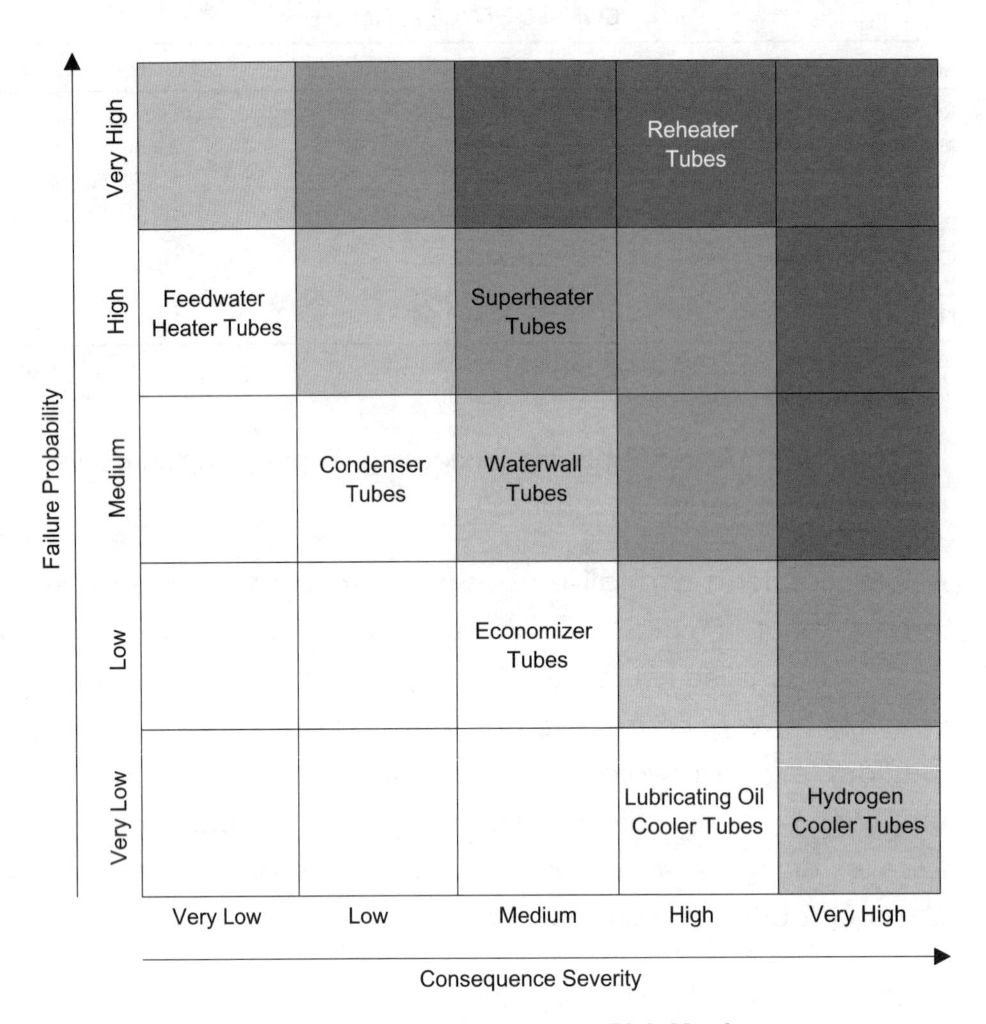

Figure 5.7.1 Tubular Component Risk Matrix

- Hydrogen cooler tube leaks could possibly introduce water to the generator or cause an explosive atmosphere.

Again, we have identified five fairly clear consequence levels, which are identified in Table 5.7.1. If we had not been able to rank the components, we could have interviewed the operators again.

This data produces the risk matrix in Figure 5.7.1. Recall the following from the previous discussion about the risk matrix:

1. Risk increases on a diagonal, upward to the right.
2. The three levels of shading therefore represent "bands" of more-or-less equal risk.

3. Reheater tubes clearly need the most immediate attention, followed by superheater tubes.
4. Hydrogen cooler tubes are worth a closer look because of their consequence rating.

Observe that value has been added even at this early stage of risk-based program development. Several forced outage contributors have been logically prioritized in an easily-communicated way.

NOTES:

SECTION 6

QUANTITATIVE RISK ANALYSIS

To perform this procedure, you will need:

- *The identified system of interest, broken down into trains (if applicable) and components. See Section 3.*
- *(Optional) Facility-specific data. See Section 4.2.*
- *(Optional) A preliminary qualitative risk ranking. See Section 5.*

This procedure will produce:

- *Incipient understanding of quantitative risk-based methods.*

The next step will, in general, be quantitative risk assessment for the selected components, starting with generic data analysis. If the generic data is inadequate, you might develop and combine data from other sources. If you are working with a complex system, you will also need fault tree/event tree analysis. You might want to reconsider system information assembly.

QUANTITATIVE RISK ANALYSIS 6

The qualitative risk assessment ranked risk in terms of broad failure probability and consequence bands. This much detail is sometimes enough, however, as was noted in Section 5, qualitative risk measures become less valuable as system complexity increases, as analysis scope broadens and as the need arises to compare risk among facilities, particularly if the facilities are different. For example, is a high probability/medium consequence event in a coal boiler more or less severe than a same-rated event in a similar boiler at a different station? Is it more or less severe than a same-rated event in a gas turbine? Quantitative risk analysis answers these questions.

This Handbook explains how to schedule component replacements and how to design effective inspection programs. These tasks require quantitative risk analysis. However, even if you intend to push the risk-based methodology as far as it will go, qualitative risk assessment is a logical first step, because its results can serve as a reality check during some of the quantitative procedures. A qualitative risk assessment can also serve as a risk-based tool that can screen out components of less concern early in the process.

Quantitative risk analysis assigns numerical values to failure probabilities and consequence costs. You obtain these numbers from various sources, probably including generic databases, elicited expert opinion, your plant's management, load dispatcher, systems planning department, financial department, archives and a variety of other sources. The first key product is a failure probability versus time curve for each component. For a complex system, however, component data is not immediately useful for determining system effects, so you might use fault tree or event tree analysis to explore the relationship between component and system performance.

Up to this point in the analysis, you are dealing only with the immediate component failure consequences. This is sufficient, perhaps, for the initial projects that correct obvious, relatively small issues, but several factors conspire to require a deeper look.

First, you don't get very far into even the "Pareto 20%" before the number and size of the attractive projects you can identify will overwhelm any real-world maintenance and capitol-improvement budget. "Pareto 20%" refers to a thumb rule that is based upon the Pareto Principle. Pareto discovered that 20% of the population produced 80% of the wealth. This principle is frequently applied in management, e.g., 20% of the items cause 80% of the problems. It is also used in Total Quality Management work.

Next, the components are aging, each in their own way and at their own rate. This means that some components can wait longer than others can before they violate a safety or other constraint; further, variation of the real cost with time interacts with the failure probability variation with time in a way that could make the project ranking unclear.

Finally, possible variations in unit mission(s) with time add another dimension of complexity. Optimized scheduling and strategic planning are the only way to ensure that maintenance management of expensive corporate assets maximizes the NPV associated with those assets. The unique ability to consider the time-varying nature of risk is a key feature of and a powerful reason for quantitative assessment.

Once you have established component failure probability versus time and have correlated this information with the system or facility mission, you will have a facility model that can be adapted and queried to help you design inspection and component replacement programs. These programs will be designed to ensure that safety, unit availability and other constraints that you select are met during a process that will show you how to maximize probable NPV.

For several reasons, you might need a more refined answer than the data that was available during your first "pass" can produce. For example:

- You might lack data for some component(s) of interest

- Sensitivity analysis might suggest that poorly supported data is important
- Your analysis results might disagree with stakeholder opinions or your own intuition

Using more sophisticated engineering tools to reduce uncertainty is one way to refine or confirm your answer. Section 9.4 shows you how to incorporate results that you might obtain from an engineering analysis tool such as remaining life analysis. Other tools are available.

Section 9.3 shows you how to obtain expert opinion, how to convert that opinion to a failure probability vs. time curve and how to combine the curve with curves that are derived form other data sources. Once you understand the analysis procedures, you will know how to frame an inquiry for needed information and how to write a specification for obtaining needed information from a consultant.

Finally, you can use the human factors analysis tools to refine failure estimates for scenarios that involve human interactions with the system(s) you are analyzing. See Appendix D for more information.

NOTES:

SECTION 7

FAILURE DATA ANALYSIS

To perform this procedure, you will need:

- *The system of interest, broken down into trains (if applicable) and components. See section 3.*
- *Facility-specific component data. See Section 4.2, and/or*
- *Generic data. See Section 4.3.*
- *MS Excel and the Handbook CD-ROM.*

This procedure will produce:

- *Risk ranking tables*
- *Risk plots*
- *Component failure probability versus time curves*

The next step will, in general, be multiple component ranking, however, you might need to obtain and combine data from other sources. If you are working with a complex system, you will also need fault tree/event tree analysis.

FAILURE DATA ANALYSIS 7

This section receives facility-specific or generic data from Section 4.2 or 4.3 and processes it into quantitative risk plots and failure probability vs. time curves. You should not mix data from different sources. Keep the data from different sources separate. After you have generated separate failure probability vs. time curves, you can use the procedures in Section 9 to combine them.

7.1 □ DETERMINING FAILURE PROBABILITY AND CONSEQUENCES

This procedure uses the spreadsheet RISKRANK.XLS to sort and analyze forced outage (failure) data. The data can be facility-specific or generic. If you are using pc-GAR and the NERC-GADS generic database, use RISKCAUS.XLS to sort and analyze the generic forced outage (failure) data. Read Appendix B before you run any of the Handbook spreadsheets.

The spreadsheet already contains sample data that is identified by NERC-GADS cause codes. These cause codes in fact represent specific components. You should work through the procedure with the data that is already there, then over write the data with your own data. When you insert your own data, you will, of course, use your own component identifiers, if applicable, instead of NERC-GADS cause codes.

First, you will import the data into the spreadsheet. Then, you will sort the data and aggregate the number of failures and the production (MWH) loss per component. Finally, you will calculate component risk and risk-rank the components. Proceed as follows:

1. Ensure that each record (row) in the forced outage (failure) data set that you wish to use contains the following information:
   a) Component identifier or cause code
   b) Operation age (yr.) of the component at failure

c) Number of failures in each operating year

d) Maximum unit dependable production (MW or other suitable unit) during the operating year. (This field is not included if you are using the NERC-GADS risk tank spreadsheet RISKCAUS.XLS.)

e) Average Maximum Dependable Equivalent Time Duration (hr) of Lost Production (or total time duration of lost production (hr) for RISKCAUS) caused by failures of the component during the year. It is the sum of the "FO" row (Forced Outage) and "FDe" row (forced Derating) for the calendar year in the pc-GAR output. The equivalent time duration is the full forced outage hours, plus (Derate Hours* (Derate Capacity/Max. Dependable Capacity)).

f) Magnitude of production-time loss (MWH) for the operating year for RISKCAUS. It is the "FO" row plus the "FDe" row for the calendar year in the pc-GAR output.

2. Start Microsoft Excel and open RISKRANK.XLS (or RISKCAUS.XLS if you are using NERC-GADS data from pc-GAR). You will see that the spreadsheet (Raw data} tabs already contain example data for boiler tubes.
3. Either hand-enter the data into spreadsheet columns A through E or [Edit] [Copy] and [Edit] [Paste Special] [Values] the data from where you had formatted it. See Figure 7.1.1. The MWH Loss will be calculated automatically in column F. (Of course, you are entering it directly in the RISKCAUS.XLS spreadsheet.)
4. After you have loaded the data into the spreadsheet, save the spreadsheet to your working directory under a new name so that you do not lose your work.
5. From the upper left-hand corner of the spreadsheet data, drag-select all the columns and rows of input data including column F. Include the header row (row 2) and all the rows that contain data.
6. Select [Data] [Sort].
7. Select [Component Identifier] as the primary sort field and select [Ascending Order].
8. Select [Operation Age] as the secondary sort field and select [Ascending Order].

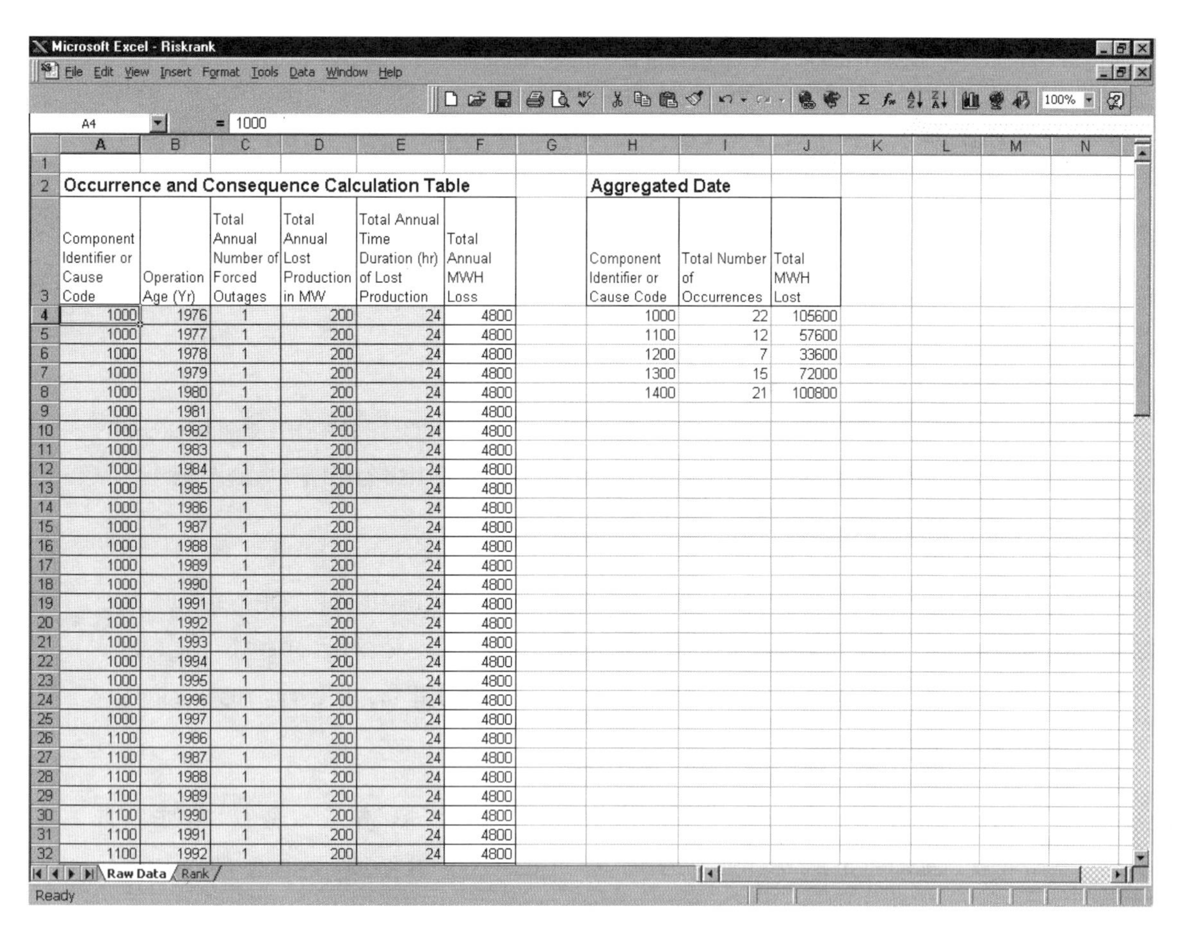

Microsoft Excel - Riskrank

File Edit View Insert Format Tools Data Window Help

A4 = 1000

Occurrence and Consequence Calculation Table

Component Identifier or Cause Code	Operation Age (Yr)	Total Annual Number of Forced Outages	Total Annual Lost Production in MW	Total Annual Time Duration (hr) of Lost Production	Total Annual MWH Loss
1000	1976	1	200	24	4800
1000	1977	1	200	24	4800
1000	1978	1	200	24	4800
1000	1979	1	200	24	4800
1000	1980	1	200	24	4800
1000	1981	1	200	24	4800
1000	1982	1	200	24	4800
1000	1983	1	200	24	4800
1000	1984	1	200	24	4800
1000	1985	1	200	24	4800
1000	1986	1	200	24	4800
1000	1987	1	200	24	4800
1000	1988	1	200	24	4800
1000	1989	1	200	24	4800
1000	1990	1	200	24	4800
1000	1991	1	200	24	4800
1000	1992	1	200	24	4800
1000	1993	1	200	24	4800
1000	1994	1	200	24	4800
1000	1995	1	200	24	4800
1000	1996	1	200	24	4800
1000	1997	1	200	24	4800
1100	1986	1	200	24	4800
1100	1987	1	200	24	4800
1100	1988	1	200	24	4800
1100	1989	1	200	24	4800
1100	1990	1	200	24	4800
1100	1991	1	200	24	4800
1100	1992	1	200	24	4800

Aggregated Date

Component Identifier or Cause Code	Total Number of Occurrences	Total MWH Lost
1000	22	105600
1100	12	57600
1200	7	33600
1300	15	72000
1400	21	100800

Raw Data / Rank

Ready

Figure 7.1.1 [Raw data] in RISKRANK.XLS

9. Select (my list has) [header row].
10. Click on [OK] and the sort will run.
11. Save the file to your working directory under a suitable name. It contains your failure data, sorted by component identifier (cause code). You will need it later.
12. Select [Tools] [Aggregate] to run the macro that will aggregate the data for each cause code into a total number of failure occurrences and total production (MWH) lost.
13. Now copy the rows under the heading "Aggregated Data" that contain component identifiers (cause codes), total numbers of failure occurrences and total production (MWH) losses into the left labeled three columns in the [Rank] tab
14. Note that, as you enter the data, the spreadsheet calculates, in the

fourth column, the total risk for each component identifier (cause code), i.e., the product of the total number of occurrences and the total production (MWH) loss for each component identifier (cause code).

15. Save this file to your working directory for future use. It is called the aggregation spreadsheet.

7.2 □ RISK ASSESSMENT

In this section, you will perform a quantitative risk assessment and screening. You can use either a risk-ranking table or a risk plot (or both!).

7.2.1 □ Assessment by Risk-Ranking Table

Now that you have accumulated the components of concern in the aggregation spreadsheet, it is time to rank all the components that you have selected. We'll risk-rank the list by sorting it by the number of occurrences and MWH loss.

1. Under the [Rank] tab in RISKRANK.XLS or RISKCAUS.XLS, drag-select the header row (row 2) and all the rows of columns A through D that contain data.
2. Select [Data] [Sort].
3. Select [Risk] as the primary sort field, [Descending Order] and (my list has) [header row].
4. Click on [OK] and the sort will run.
5. Save the resulting file. It will contain a sorted list of risk-ranked components of concern.

The list still contains too many components to address at one time, so we will weed out the less significant ones.

1. Inspect the table that you have created. There are probably a number of components (cause codes) at the bottom of the ranked table that have much lower risk levels than those that proceed them in the list. In the example, cause codes 1200, 1100 and (arguably) 1300 fit this description. You will want to delete these components and focus your attention on the remaining (most important) items.
2. If, after you drop obviously lower-risk items, there are still too many

components (cause codes) to analyze on the first pass, then select the top 20% of the components for the first pass and delete the rest. The 20% criterion, which is from Pareto was introduced in Section 6. Apply this thumb rule by counting the components (cause codes) and deleting columns A through C for all except the "top 20%." In the example, cause code 1000 is the only cause code you would retain by this criterion. If this was real data, however, you would probably want to retain cause code 1400 as well. The risk rankings are based upon a very limited sample, and the risk values for cause codes 1000 and 1400 are so close that a very small change in the data would reverse them in the ranking.

3. Save the spreadsheet to your working directory. It contains a prioritized (and culled) risk-ranking table.

Do not be concerned about the components that you are dropping. You will reexamine those components after you have prioritized the selected "first pass" components and developed an inspection program for them.

7.2.2 □ Assessment by Risk Plot

First, we will need to plot the number of occurrences against the production (MWH) loss for every component. Then we will choose the high-risk components.

1. Open the spreadsheet template RISKPLOT.XLS from the CD-ROM. Save it to your working directory under an appropriate name. See Figure 7.2.2.1.
2. In the same Excel session, open the previously aggregated and sorted spreadsheet (originally RISKRANK.XLS or RISKCAUS.XLS) (Section 7.1). Drag-select and [COPY], from the [Rank] tab, all the cells that contain data and risk. Do not include the header row (row 2).
3. Switch to the other window (RISKPLOT copy) (⟨ctrl-F6⟩ will do this). Perform an [Edit][Paste Special][Values] into cell A3 of this spreadsheet. Note that a risk plot appears, with the Total Number of Occurrences on the vertical axis and the Total Consequences on the horizontal axis. In order to place the correct labels by the data points

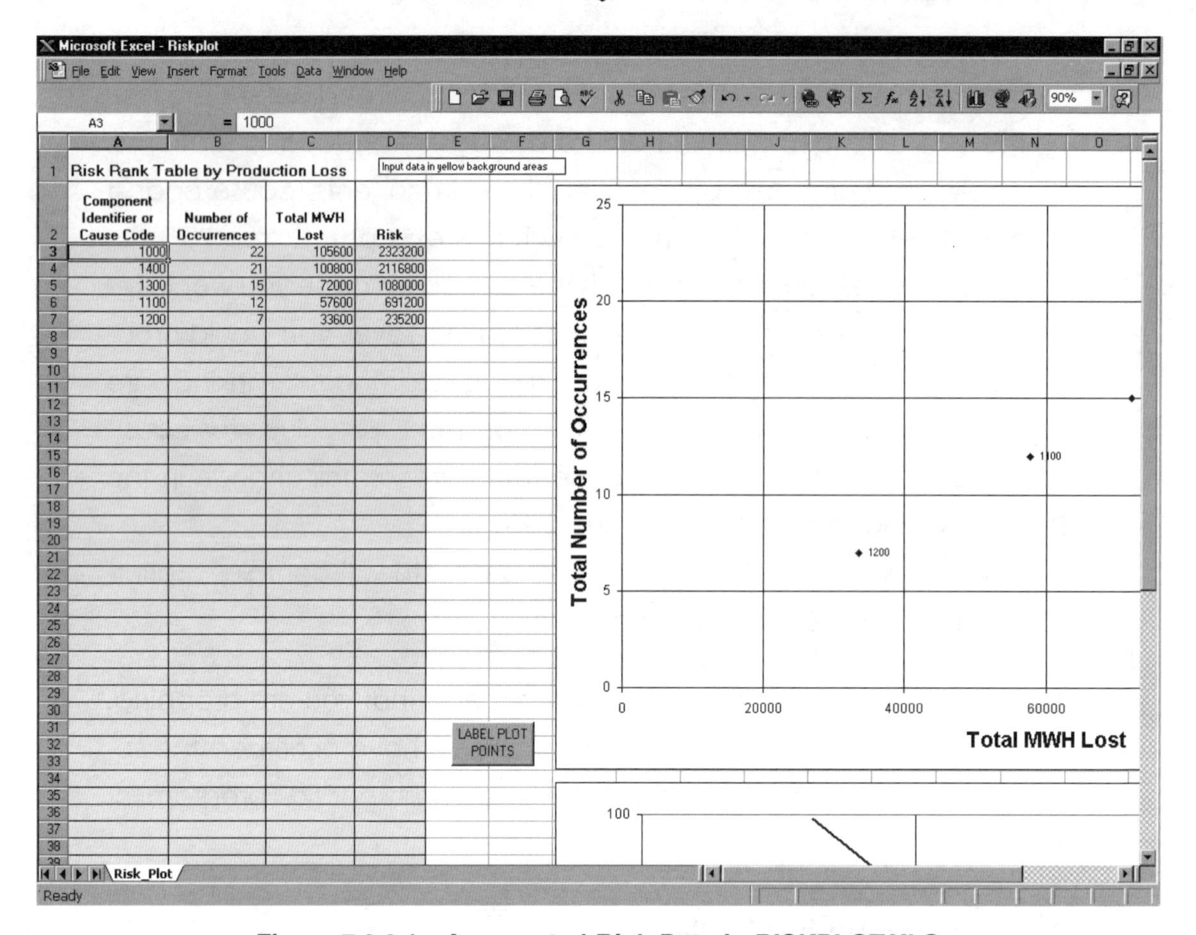

Figure 7.2.2.1 Aggregated Risk Data in RISKPLOT.XLS

from column A, select the chart by clicking on it and then click on the button titled [LABEL PLOT POINTS]. (In the example, the horizontal axis is labeled "Total MWH Lost." When you customize the spreadsheet, you can change the axis to a different measure of production loss if necessary.)

4. A log-log plot also appears below the previously-described plot. In order to place the correct labels by the data points from column A, select the chart by clicking on it and then click on the button titled [LABEL PLOT POINTS]. The log-log plot shows the same axes and also a line of constant risk. You can place the line anywhere on the risk plot by selecting the plot, clicking on the line and moving the ends of the line with the left mouse button.

   If, after you have placed the line, the product of a point on the vertical axis with its corresponding point on the horizontal axis is equal to the product of a point on the horizontal axis and its corresponding point on the vertical axis, then you have a line of constant risk when these end points are joined with the line. The easiest way to place the line is to choose, by engineering judgment, a number of occurrences and/or total consequence that seems excessive at this pass.

5. In the example data, the cause codes 1000 and 1400 are extremely close together, and the other cause codes are of significantly lower risk.
6. Save the file to your working directory. It contains the divided risk plot.

The line that you have just placed is the minimum risk line. The components (cause codes) that are to the right of this line identify the components that you will carry into inspection prioritization and program development during the next steps. Those to the left of the line you can ignore until you are choosing a set of components for the next pass.

7.3 □ ESTIMATING COMPONENT FAILURE PROBABILITY VERSUS TIME

This procedure uses the spreadsheets PROBCALC.XLS and BAYCOM-11.XLS to process failure data and produce failure probability vs. time curves. The data can be facility-specific or generic.

PROBCALC.XLS already contains sample data that is identified by NERC-GADS cause codes. These cause codes in fact represent specific components. You should work through the procedure with the data that is already there, then over write the data with your own data. You will, of course, use your own component identifiers, if applicable, instead NERC-GADS cause codes.

If you are using NERC-GADS data from the pc-GAR program that is described in Section 4, then use PROBCAUS.XLS instead of PROBCALC.XLS to determine the probability of failure versus time from the forced outage data. See Section 7.1.

1. Ensure that each record (row) in the forced outage (failure) data set that you wish to use contains the following information:
   a) Component identifier or cause code

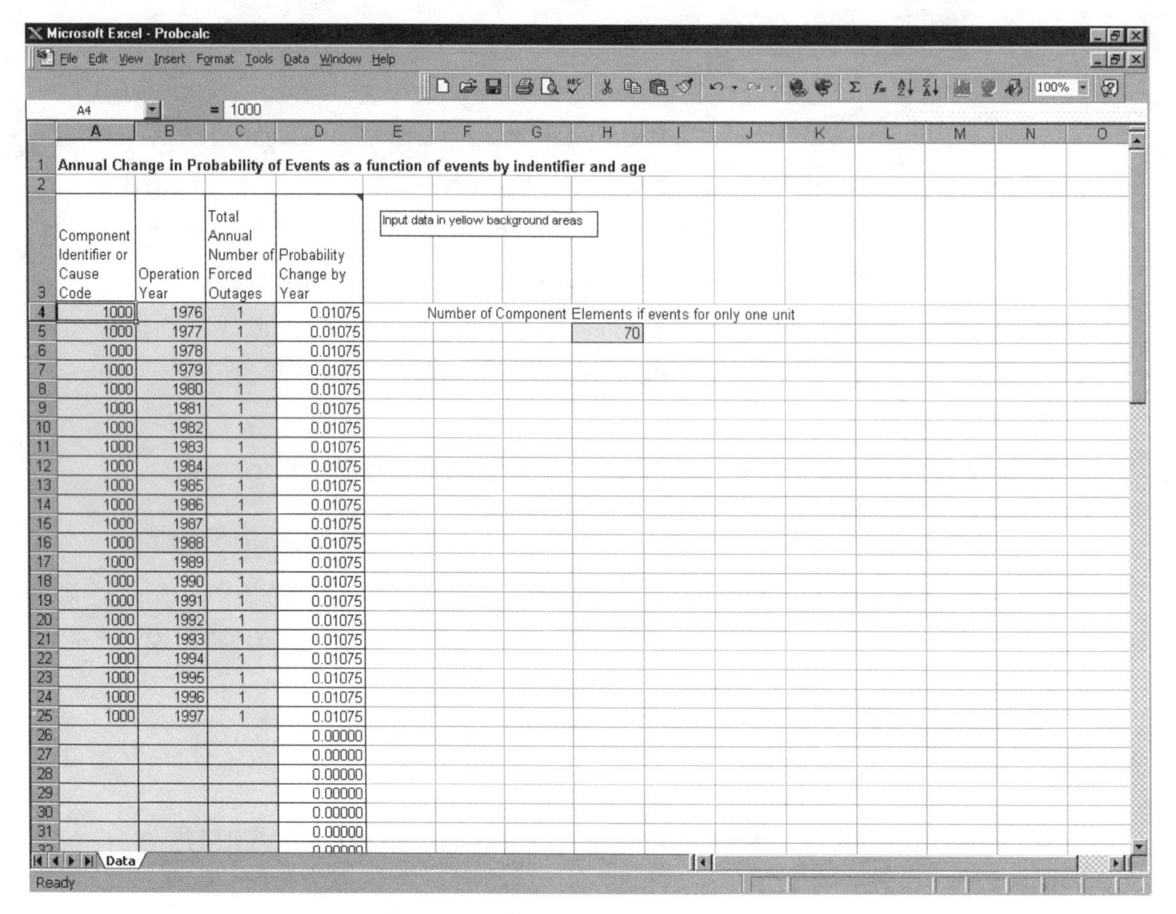

Component Identifier or Cause Code	Operation Year	Total Annual Number of Forced Outages	Probability Change by Year
1000	1976	1	0.01075
1000	1977	1	0.01075
1000	1978	1	0.01075
1000	1979	1	0.01075
1000	1980	1	0.01075
1000	1981	1	0.01075
1000	1982	1	0.01075
1000	1983	1	0.01075
1000	1984	1	0.01075
1000	1985	1	0.01075
1000	1986	1	0.01075
1000	1987	1	0.01075
1000	1988	1	0.01075
1000	1989	1	0.01075
1000	1990	1	0.01075
1000	1991	1	0.01075
1000	1992	1	0.01075
1000	1993	1	0.01075
1000	1994	1	0.01075
1000	1995	1	0.01075
1000	1996	1	0.01075
1000	1997	1	0.01075
			0.00000
			0.00000
			0.00000
			0.00000
			0.00000
			0.00000

Figure 7.3.1 Individual Cause Code Forced Outage Data in PROBCALC.XLS

b) Operation age (yr.) of the component at failure

c) Number of failures in each operating year

The spreadsheet that you saved in Section 7.1 step 10 will serve. If you wish, you can use data that you did not process through risk ranking. Follow Section 7.1 steps 1 through 10 to prepare the data.

2. Open the example spreadsheet PROBCALC.XLS that is located on the Handbook CD-ROM. Save it to your working directory under a suitable name. See Figure 7.3.1.
3. In the same Excel session, open the spreadsheet that contains your data.
4. In your data spreadsheet, select the data for the component (cause code) of interest by drag-selecting columns A through C from all the rows of data that contain the cause code of interest. You should start

with the first component in the risk-ranking table or the upper right-most component in the risk plot. Do not include the header row.

5. Select [EDIT] [COPY].
6. Switch to the other window (PROBCALC copy) (⟨ctrl-F6⟩ will do this). Perform an [Edit][Paste Special][Values] into cell A4 of this spreadsheet
7. Now call the macro that will calculate the failure probability for each line of the spreadsheet by selecting [Tools] [Probability Calculation]. You will see the changes in forced outage probability for this cause code for the time represented by this line in the spreadsheet.
8. Save the spreadsheet to your working directory with a name that contains the component identifier (cause code).
9. Repeat steps 1–6 until you have estimated probabilities for all the cause codes that you want to consider at this time.

These spreadsheets are the basis for the failure probability curves that you will derive from the failure history or generic data.

A macro in the BAYCOM11.XLS spreadsheet will fit your selected failure history data with a three-parameter Weibull distribution equation. We use this distribution because it was originally derived for failure history data and because it is very robust in its ability to fit complex data. Note the Alpha, or shape parameter; Beta, or scale parameter; and base year values that were produced from the failure history data in the example. The distribution equation that these parameters define will predict the component failure history during later calculations.

As your analysis progresses from failure history through data combination; risk calculation, ranking and screening; and finally maintenance action timing optimization, the distribution equation carries probability information from one step to another. The Alpha parameter controls the general shape of the curve and is very sensitive. The Beta parameter controls both the probability and time scaling. The base year, which is usually the year of component installation or the last year that the component was replaced, locates the curve in time. You must subtract the base year from the current year because the Weibull distribution equation fits the component failure data reasonably well only during the lower tens of years of

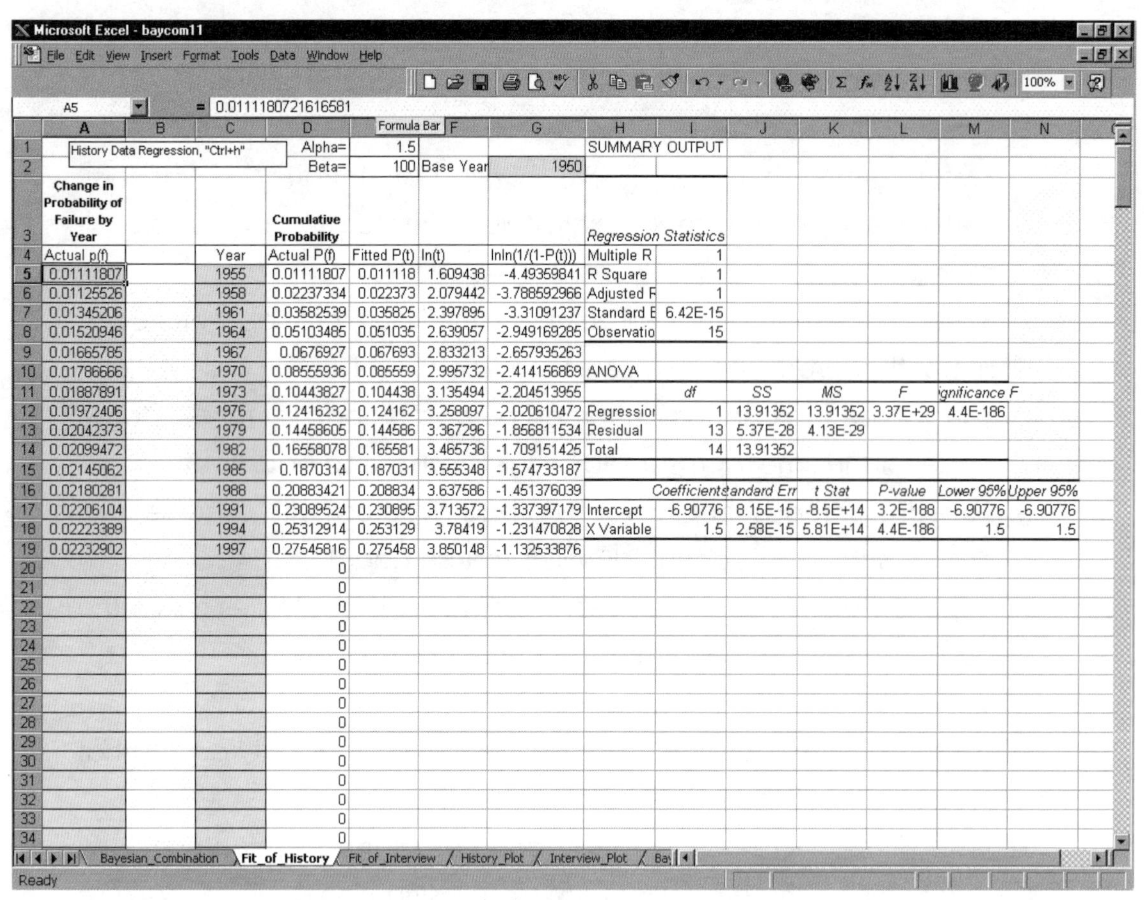

	A	B	C	D	E	F	G	H	I	J	K	L	M	N
1	History Data Regression, "Ctrl+h"			Alpha=	1.5			SUMMARY OUTPUT						
2				Beta=	100	Base Year	1950							
3	Change in Probability of Failure by Year			Cumulative Probability				Regression Statistics						
4	Actual p(f)		Year	Actual P(f)	Fitted P(t)	ln(t)	lnln(1/(1-P(t)))	Multiple R	1					
5	0.01111807		1955	0.01111807	0.011118	1.609438	-4.49359841	R Square	1					
6	0.01125526		1958	0.02237334	0.022373	2.079442	-3.788592966	Adjusted R	1					
7	0.01345206		1961	0.03582539	0.035825	2.397895	-3.31091237	Standard E	6.42E-15					
8	0.01520946		1964	0.05103485	0.051035	2.639057	-2.949169285	Observatio	15					
9	0.01665785		1967	0.0676927	0.067693	2.833213	-2.657935263							
10	0.01786666		1970	0.08555936	0.085559	2.995732	-2.414156869	ANOVA						
11	0.01887891		1973	0.10443827	0.104438	3.135494	-2.204513955		df	SS	MS	F	ignificance F	
12	0.01972406		1976	0.12416232	0.124162	3.258097	-2.020610472	Regressior	1	13.91352	13.91352	3.37E+29	4.4E-186	
13	0.02042373		1979	0.14458605	0.144586	3.367296	-1.856811534	Residual	13	5.37E-28	4.13E-29			
14	0.02099472		1982	0.16558078	0.165581	3.465736	-1.709151425	Total	14	13.91352				
15	0.02145062		1985	0.1870314	0.187031	3.555348	-1.574733187							
16	0.02180281		1988	0.20883421	0.208834	3.637586	-1.451376039		Coefficients	andard Err	t Stat	P-value	Lower 95%	Upper 95%
17	0.02206104		1991	0.23089524	0.230895	3.713572	-1.337397179	Intercept	-6.90776	8.15E-15	-8.5E+14	3.2E-188	-6.90776	-6.90776
18	0.02223389		1994	0.25312914	0.253129	3.78419	-1.231470828	X Variable	1.5	2.58E-15	5.81E+14	4.4E-186	1.5	1.5
19	0.02232902		1997	0.27545816	0.275458	3.850148	-1.132533876							
20				0										
21				0										
22				0										
23				0										
24				0										
25				0										
26				0										
27				0										
28				0										
29				0										
30				0										
31				0										
32				0										
33				0										
34				0										

Figure 7.3.2 Probability of Failure Data in [Fit_of_History] Worksheet in BAYCOM11.XLS

component life.

Proceed as follows:

1. Copy the BAYCOM11.XLS example spreadsheet template from the Handbook CD-ROM to your working directory. Use a name that will be meaningful to you and the extension .xls. See Figure 7.3.2.
2. Select the [Fit_of_History] worksheet tab. You will enter data in the yellow-background areas, starting with cell A5.

Copying and Pasting Data	Entering Data Manually
3a. Open and select the previous probability calculation spreadsheet. Highlight the probability column and select [Edit] [Copy].	3b. Obtain tabulated failure probability vs. operating year data.

4a. Reselect the [Fit_of_History] tab and select cell A5. Select [Edit] [Paste Special] [Values]. Then click on [OK].

4b. Manually enter the probability of failure data, year-by-year, into column "A."

5a. Return to the probability calculation spreadsheet, select the Time data associated with the history probabilities by year and copy them into the [Fit_ of_History] tab starting at cell C5.

5b. Manually enter each associated year into column "C."

6. Type the Base Year to be used for this data fit into cell G2. The base year is defined as the year the component was replaced or the year the unit was started, which ever is later.
7. To fit a regression curve, click [Tools] [Fit History]. Cells E1, E2 and G2 will then contain the Weibull-fitted equation alpha, beta and base year respectively. Record these three values; they are the basis for the failure probability curve for this component (cause code). You will use this curve in Section 9 or Section 11.
8. Perform a [File][Save As] to your working directory and provide a new file name. You may use this spreadsheet later for entering interview data and/or for performing Bayesian Combination Analysis with other data for the component that it represents. See Section 9.
9. Once this step has been performed for the first component, go back to the risk-ranking table or risk plot, pick the next component or the next component to the right of the line and repeat the process starting with step 4 in the first procedure under Section 7.3.
10. Repeat this process until you have alpha, beta and base year values for all the components in the culled risk-ranking table, for all the components to the right of the line in the risk plot or for all components of interest.

You have now established a failure probability versus time projection curve for each component based on its failure history.

NOTES:

SECTION 8

To perform this procedure, you will need:

- *The identified system of interest, broken down into subsystems and components. See Section 3.*
- *Capacity data for redundant subsystems and components.*
- *Failure rate data for each component, developed from generic data (Section 4.3), plant-specific data (Section 4.2) or a combination (Section 9).*

This procedure will produce:

- *(Optional) A reliability block diagram.*
- *An event tree for a selected consequence.*
- *A fault tree for a selected event.*
- *(Optional) A system failure probability vs. time curve.*

The next step will, in general, be multiple component ranking based upon optimization.

SYSTEM ANALYSIS TOOLS 8

Failure rates, whether you derived them from generic data, (Section 4.2) plant specific data (Section 4.4) or a combination (Section 9), are directly useful if the components, trains (if applicable), or systems are not redundant. This is because, without redundancy, component failure directly causes system failure. With redundancy, component failure contribution to system risk is not necessarily clear; therefore, raw component failure data can misdirect your inspection or testing program. You will need additional analysis.

To obtain the relationships that you need, you can build fault trees and event trees and subsequently quantify them. Fault trees and event trees graphically model system and component interrelations. To "quantify" a tree is to assign probabilities and/or consequences to its events or "nodes" (branches). A program that will do this, Integrated Risk and Reliability Analysis System (IRRAS), is included on the Handbook CD-ROM. Its reference manual and tutorial introduce SAPHIRE, another suitable tool. The CD-ROM also contains The Fault Tree Handbook, which is a reference text for fault tree/event tree analysis. The Fault Tree Handbook, Chapter 12, lists other analysis programs. See Appendix C.

When you have modeled the redundancy and other complexities of your facility, you can exercise the model with time-variant component data to produce system, unit or facility level failure probability vs. time curves. IRRAS and similar tools can produce such information directly if you enter the basic events as probability distributions. These analysis methods are, however, outside the scope of this Handbook. Consult the IRRAS reference manual or the instructions for a similar package if you wish to pursue this methodology.

8.1 □ RELIABILITY BLOCK DIAGRAMS

To help you to understand the relationships among the plant systems and to provide a basis for communication, you can prepare a reliability block diagram like the one shown in Figure 8.1.1. To prepare a reliability block diagram for a unit, you would:

1. Identify the systems that are required to produce output, such as the fuel handling system, fuel system, draft system, feedwater system.
2. Divide the systems into trains (parallel paths) to show degrees of redundancy. For example, the various parts of the fuel and combustion air systems in Figure 8.1.1 have different degrees of redundancy.
3. Identify on the diagram the amount of redundancy that is required. For example, 3 out of 4 fuel systems are required for 66% to 100% power operation.

Because low power and high power operation require different numbers of redundant components or trains, you might need separate reliability block diagrams to clearly identify which plant system combinations are needed for low and high power.

The reliability block diagram is easily understood by most maintenance, operational and management personnel. The diagram is a basis for discussions about and agreement on plant operation. By providing a familiar picture that all can understand, the diagram also provides credibility for the program.

8.2 □ EXAMPLE

Section 3.3, the system definition example, showed a schematic for one of three boiler feed pumps in a coal-fired power plant. There are three feedwater pump trains in the unit. See Figure 8.2.1.

8.3 □ EVENT TREES

An event tree is a "bottom up" graphic that can provide:

- Event probability and severity information.
- System insights that will suggest areas for improvement.

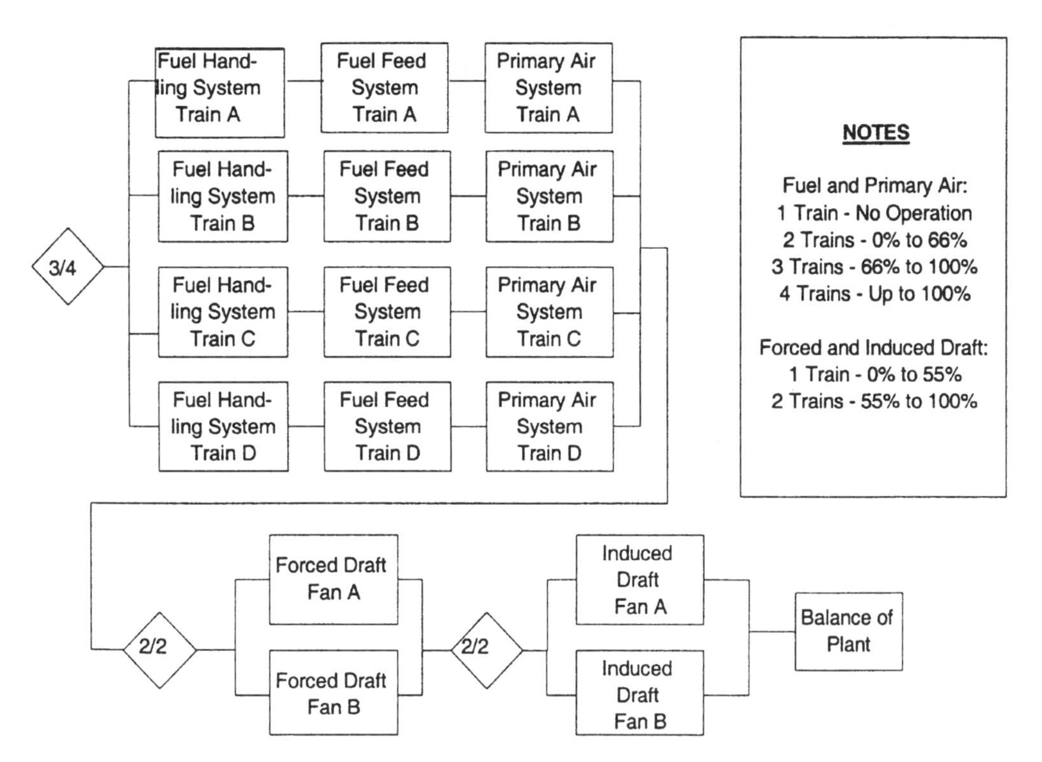

Figure 8.1.1 Reliability Block Diagram for Boiler Fuel and Combustion Air Systems

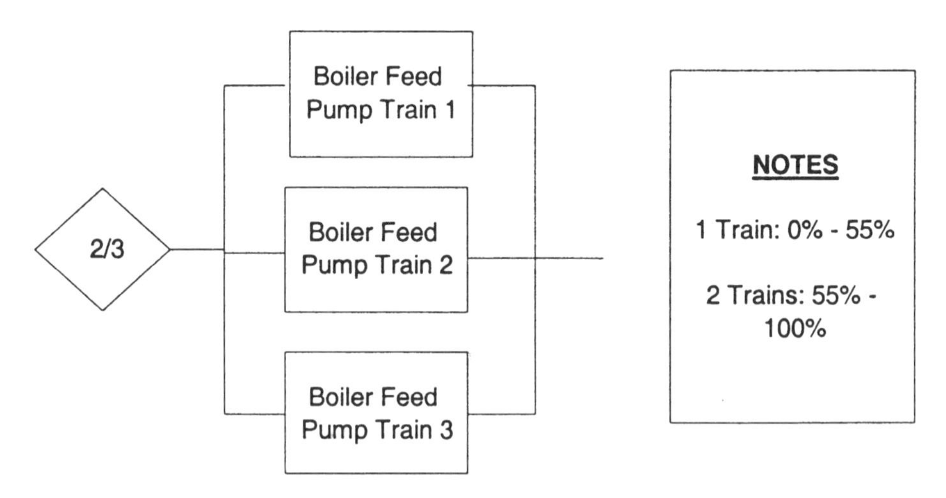

Figure 8.2.1 Reliability Block Diagram for the Main Feed Pump System

"Bottom up" means that the analysis starts with an initiating event, generally a failure. For example, you might want to consider a specific failure event, such as tube failure in a water tube boiler (Figure 8.3.1) or generator damage that is allowed by an electrical protective system

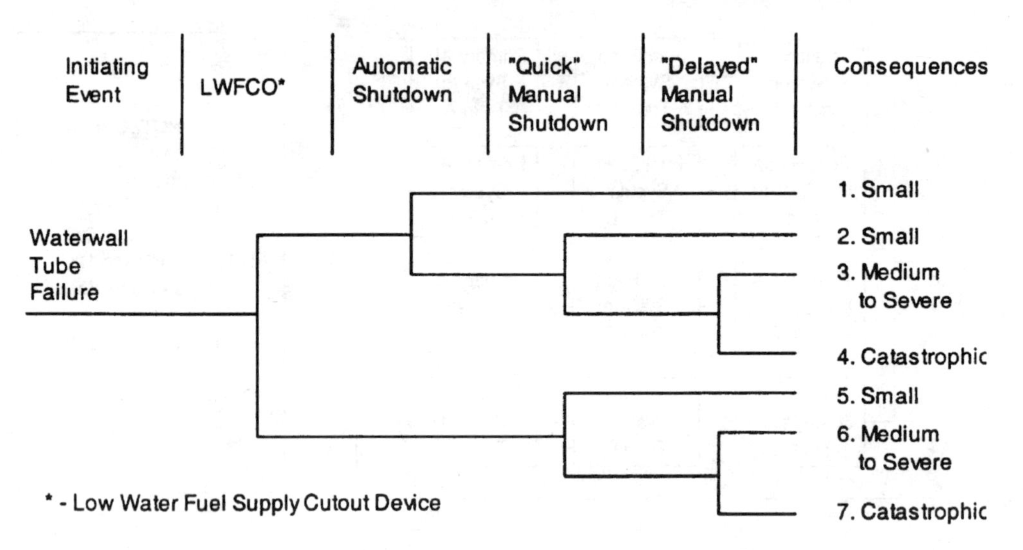

Figure 8.3.1 Event Tree for a Waterwall Tube Failure

failure. You can fully analyze these and most other failure events only by considering a failure-initiated series of events, operator actions and physical responses that result in consequences that may vary from minor to catastrophic.

An event tree is a series of paths that logically identify the results of initiating and subsequent events. Each path through the tree is called a sequence. A quantified event tree provides sets of occurrence probabilities that lead to sequence probabilities. Each sequence leads to a consequence. If you estimate or calculate the cost associated with each consequence, you can combine these values with the probabilities to obtain the risk associated with each sequence. You can do this qualitatively or quantitatively. The event tree also provides insight into the event sequences that will suggest where to make changes in equipment, controls, configuration or operator action that will either decrease the catastrophic failure probability or its consequence severity.

When you develop an event tree, you should work from left to right on the page, proceeding as follows:

1. Identify an initiating event such as a component failure, operational configuration or operator error and list it as the first item on the left.
2. To the right of the initiating event, list a responding system (e.g., low water fuel cutout (LWFCO)), operator action (e.g., manual

shutdown), control loop (e.g., automatic shutdown), environmental condition or other item that will mitigate, aggravate or propagate the event. This response will work, partially work or not work. Each option forms a "branch" of the event tree. In Figure 8.3.1, all the options, which are listed across the top of the figure, either succeed or fail. In all cases, following convention, the "up" branch represents success and the "down" branch represents failure.

3. Continue to work to the right. For each branch, list another responding system, operator action, control loop, environmental condition or other item that mitigates, aggravates or propagates the resulting plant condition. This will produce two or three branches for each previous branch.
4. Continue this process until you reach the plant condition (consequence) of concern, such as plant fire, boiler tube failure, or generator damage.
5. For each of the last branches, identify the consequence of the sequence of events that led to the branch. You will probably need to obtain the opinions of a number of experienced people in the company to develop this information. Figure 8.2.1 shows qualitative consequences.
6. To quantify the tree, for each of the event tree branch points, estimate the probability that each choice will occur. Note that the sum of the branch probabilities that lead from each branch point must be 1.
7. Multiply the probabilities for each path from initiating event to consequence to obtain the probability that the consequence will occur because of the events in the path.
8. Estimate or calculate the expected consequence cost (or risk) for each "final" plant condition. To obtain the risk associated with each path, multiply the expected costs by the probabilities that you calculated in step 7.

Fault trees, which are introduced in the next section, work with event trees when you are analyzing complex systems. For simple events, you might estimate the branch probabilities directly. For more complex events, like the probability of automatic burner shutdown in the previous example, you might use, e.g., the event "automatic burner shutdown fails" as the "top event" in a fault tree. The top event in a fault tree is the event that is to be analyzed.

You can use IRRAS to build and quantify event trees and fault trees. In fact, IRRAS is an integrated analysis package. When you use it, you must always build an event tree, even if it is a single-branch tree with the top event (failure) in the fault tree that you are building as the single event.

8.4 □ EXAMPLE

The IRRAS Tutorial, Lesson 2, provides step-by-step instructions for building a simple event tree. Appendix C tells you how to load IRRAS and how to access the IRRAS Tutorial and Reference Manual.

8.5 □ FAULT TREES

Fault tree analysis (FTA) is a powerful analysis tool in its own right, however, fault tree analysis can only provide the failure probability for the "top event" given the system configuration and the failure probabilities for the "basic events." The top event is the event that is to be analyzed. Basic events are failure-causing "subsidiary" events that have known probabilities or which are simple enough to allow reasonably accurate probability estimates. This calculated probability can be very useful for analyses that involve simple consequences, but in quantitative risk-based analysis we generally need failure probability vs. time. If FTA has been done at your facility, you might be able to use the existing trees in your program. Whether you are fortunate enough to have existing trees that you can use or you must build your own trees, you will find you have three possible outcomes:

1. You have no components of interest that have time-varying failure rates, or you (now) have neither data nor any other rational basis for estimating the time-variation. In this case, you will only need one calculation of the tree to calculate the required top event failure probability, and you will use this as a constant failure rate in other calculations.

2. You have only one component of interest that has a variable failure rate. In this case, the relationship between component failure rate and top-event failure rate will be a constant ratio, and you will only need one calculation of the tree to obtain the ratio. This is a trivial case that is very unlikely in practice, and we will say no more about it. If your situation fits this case, of course, take advantage of it.
3. You have more than one component of interest that has a variable failure rate. In this case, the relationship between component failure rate and top-event failure rate may be complex. You will need to calculate the tree once for each year for which you have component failure rate data, then convert the resulting data to a top-event failure probability vs. time curve.

For each event that you wish to model (such as "failure to maintain 100% power" or "failure of automatic burner shutdown system"), you will:

1. Use fault trees to identify what incidents ("basic events") or incident combinations will produce the top event if they occur
2. Put the failure rate or incident occurrence rates into the fault tree
3. Quantify the fault tree
4. (Optional) Exercise the fault tree to produce time-related failure probability curves for the top event.

A quantified fault tree provides:

- The frequency with which the top event will occur, for example, "unavailability to produce 100% power."
- A list of the basic events and the groups of basic events that could cause the top event. Each such event or event group is called a **cut set**. Cut sets with only one component in them are generally the most significant.
- The contribution of each basic event and group of basic events to the top event.

Fault trees are drawn using standard symbols. See Appendix C.1 Table 4-1 or Appendix C.3 Figure 21. These symbols represent various kinds of events and the logical paths or "gates" that connect the events.

The Fault Tree Handbook describes fault trees, their theory, elements, construction and evaluation in some detail. We encourage you to read this book for understanding, however, manual fault tree evaluation can be a very tedious process. Therefore, if you are going to use fault trees, you should also become familiar with IRRAS or a similar program. See Appendix C for access to both these resources.

If you produced a reliability block diagram, it provides the top event(s) of the fault tree(s), because failure of each block(s) will prevent accomplishing the plant mission. You can expand the top event that each block represents as needed until you reach the basic event level. Figure 8.5.2 is a fault tree that was developed from the reliability block diagram in Figure 8.2.1 and the system schematic in Figure 8.5.1.

As an example of how a physical system could be translated to fault tree logic, consider the schematic in Figure 8.5.1. Figure 8.5.2, the fault tree, models the system as follows:

- The basic event "Loss of power/breaker failure" represents the electrical supply to the pump.
- "BFP operator switching error" represents motor controller, breaker and electrical bus maloperation.
- "BFP fails to continue running" represents motor, pump and coupling component failures.
- Failure of the check valve to remain open, failure of either/both isolation valves to remain open, speed control coupling failure and local loss of control air are each represented by basic events.
- Control system failures are expanded (in a separate tree that is not shown here) below the gate "BFP speed control system fails."

Some of the event choices are not obvious. "BFP fails to continue running," for example, was chosen because the failure data would not support separate failure probabilities for the motor, pump, and mechanical connections. This event would be an obvious candidate for data refinement if it proved to be important. It is not very likely to be important, if for no other reason then the fact that at least two trains must fail for the system to fail. (See the reliability block diagram, Figure 8.2.1.) Therefore, although it might look easy to refine this event, any effort to do so could

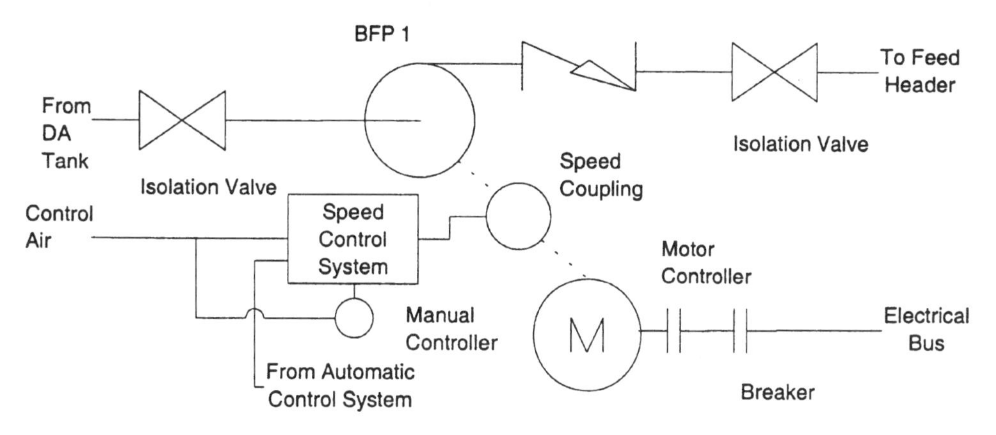

Figure 8.5.1 Boiler Feed Pump (BFP) Train Schematic (1 of 3)

probably be better invested elsewhere, e.g., on events that are part of single component cut sets.

This fault tree is part of the fault tree that is discussed in Section 8.6. Most of the gates in this tree are "OR" gates, for example, "BFP train 2 fails." An OR gate shows that the event above it fails if any of the events that directly "feed" it occur. The top gate, "BFP system fails," is a special case of the OR gate, called an n/m gate. It means that the top event in this tree fails if 2 out of 3 of the events that feed the top gate fail. "BFP control system fails is a "transfer" gate that shows that the logic continues in another tree. All the other symbols, such as "Loss of control air" are basic events.

Assuming that you went to the trouble of building a fault tree because you wanted to do a risk-based maintenance optimization or more, you will need a unit or facility failure probability vs. time curve that is based upon the component-level data. One way to obtain such a curve is to calculate the fault tree for each year for which you have component-level failure probability data. This will give you a year-by-year series of top-event failure probability data points.

To use fault tree analysis in risk-based analysis, proceed as follows:

1. Identify an event tree sequence (see Section 8.3) that has system success as an outcome. If you are going to use IRRAS, you must construct at least a single-branch event tree in which success or failure of your top event produces system success or failure sequences. If

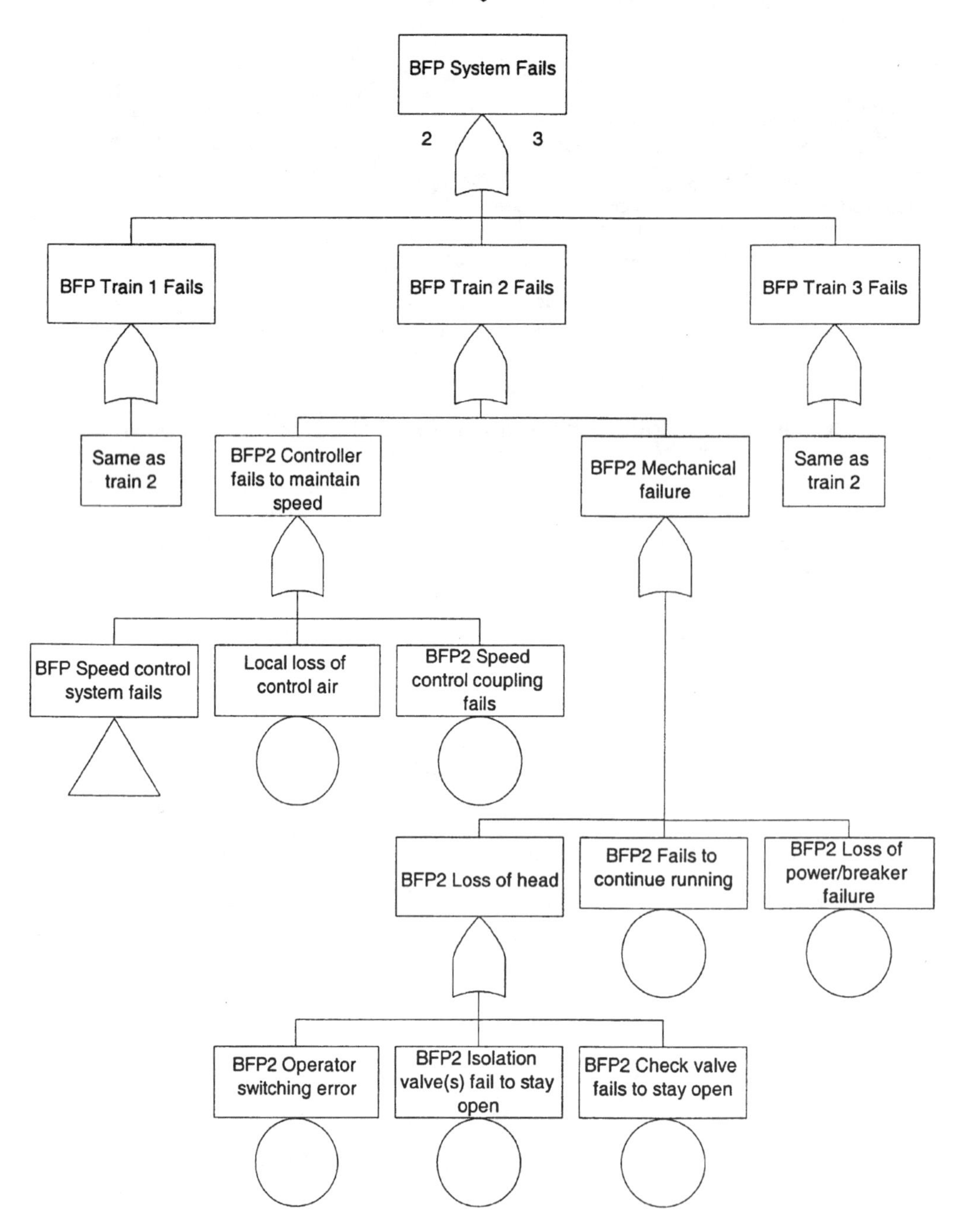

Figure 8.5.2 Boiler Feed Pump (BFP) System Fault Tree

your system is complex, you might need a more complex event tree and several fault trees to model it.

2. Assign probabilities to each gate in the event tree for which probabilities are known.
3. Build fault trees as needed to bridge the gap between your lower or

component-level failure rate data and the event tree gate events. Every component of interest must be a "basic event." Other components can be "lumped" as desired or as the available data require or allow.

4. Enter your year-by-year failure rate data and calculate the tree(s) for each year. Record the system failure probability for each year.
5. Go to Section 9.4.3 step 15 and follow the instructions for converting engineering analysis data into a failure probability vs. time curve.

8.6 □ EXAMPLES

The IRRAS Tutorial, Lessons 3 and following, provides step-by-step instructions for building simple fault trees. Appendix C tells you how to load IRRAS and how to access the IRRAS Tutorial and Reference Manual.

The /EXAMPLES/FTREE directory on the Handbook CD-ROM also contains a fault tree for a complete coal-fired generating unit. To view and analyze the tree, follow the instructions in the readme.txt file and the IRRAS Reference Manual (Appendix C.4). A few comments:

The example facility contains systems, such as Failure of Slag Removal System, in which the analyst had no great interest. It also contains systems or components, such as Failure of Attemperators, for which no failure data could be found without expending excessive resources. These systems/components were "lumped," an estimated failure rate was provided, and no further fault trees were developed. The rest of the systems and components were made top events of second level fault trees.

For the other systems, the analyst developed lower level fault trees until the needed detail and/or the limit of available failure data was reached.

8.7 □ RISK RANKING

After you have built and quantified the needed fault trees, you can use the methods that are discussed in Section 9.8 to perform a risk ranking.

NOTES:

SECTION 9

To perform this procedure, you will need:

- *A first-approximation failure probability versus time curve for one or more components of interest. See Section 7.3.*
- *Access to persons who are intimately familiar with the component(s).*
- *Design data and operating parameters for the component(s).*
- *Other valid failure rate data sources, for example, plant-specific data, generic datasets, engineering data sources and expert opinions.*

This procedure will produce:

- *A method for obtaining expert opinion in a useable form.*
- *Formatting instructions for data from various sources.*
- *An estimated component failure probability vs. time curve that is based upon formatted data.*
- *Failure probability vs. time curve(s) that validly combine data from two or more sources.*
- *Consequence of failure estimate(s).*
- *A risk ranking for components that have been taken through the data combination process.*

The next step will, in general, be multiple component ranking based upon optimization.

OBTAINING AND COMBINING DATA 9

The techniques that are described in this section use probabilistic logic to combine available data from multiple sources or data from multiple technical points of view, i.e.:

- Failure history
- Knowledgeable opinions of those who are technically involved with components day to day and/or
- Knowledgeable opinions of those who have studied the effect of the failure mechanisms that are involved in the components' performance degradation.

The method that we will use to combine the data, a Bayesian-like combination analysis, gives most weight to the data in which we have the strongest belief or the most confidence. This approach follows the philosophy that performance projections are strengthened by multiple technical viewpoints.

This section also describes how to obtain and format information from the previously-listed sources. The section ends with the risk assessment process for data that was developed and/or combined using the other processes in the section.

9.1 □ TYPES OF DATA AVAILABLE

Table 9.1 lists the data types that might be available for a plant component. The list is a good checklist to help you to consider all of the data resources that might be available.

The most commonly available component data are expert opinion, inspection results (failure history) and engineering studies. These three types of data broadly cover the resources generally available from com-

TABLE 9.1 POSSIBLE DATA RESOURCES (ADAPTED FROM ASME, 1991)

Data Source	Nature of Data	Typical Information
Design information	Specific	System and component information, functional requirements, intended loadings; can be studied to locate potential sites for various failure mechanisms
Operating experience databases	Generic	Identify failures modes and effects experienced by others; can be studied and possibly extrapolated for failure rate and failure rate vs. time information
Expert opinion	Specific	See Section 9.3.1
Engineering studies	Generic or specific	Define potential failure modes and causes. (Includes RLA results in-house and by others) See Section 9.4
Inspection results	Specific	Define potential failure modes and causes; update state of knowledge

pany personnel who are involved in estimating component life. These people are:

- Operations and maintenance personnel, who deal with the component day to day .
- Maintenance and maintenance engineering personnel, who track component performance.
- Reliability engineers and life extension or remaining life analysts, who use the latest technology to model the expected component behavior.

9.2 □ FAILURE HISTORY DATA

The best component history failure data is the failure history of the component you are studying. If you have failure history information for your component and have not yet analyzed it, follow the procedures in Sections 4.1 and 4.4.

The next-best failure history data is the failure history of a similar component in the same unit or in a similar unit. This data is still somewhat component-specific and is therefore not very heavily biased toward the worst performers in the fleet. This assumes that the data are for a truly similar component, e.g., a component of similar design with a similar service history. Data for even approximately similar components are better than generic data. There is one exception to this paragraph: if you have reason to suspect or believe that your component(s) are more harshly treated than the average. If your failure rates consistently calculate higher than published rates, you might want to compare notes with experts from other facilities. If you want to use similar-unit failure history data, follow the procedures in Section 4.2.

Finally, if you are uncomfortable with your data and/or the results it is producing and there is no specific or similar-unit data available, you will want to consider NERC-GADS or other generic data as a way to bias your data towards the experiences of a wider population. See Section 4.3 for a more complete explanation and instructions for obtaining and formatting generic data.

After you have obtained failure history data and processed it according to the applicable procedure(s) in Section 4, proceed to Section 7 to obtain a failure probability vs. time curve. Then return to this section and proceed with combining that failure history data with the other data types.

9.3 □ PLANT OPERATIONS/MAINTENANCE PERSONNEL OPINIONS

This section explains how to obtain expert opinions in a way that will be as objective as possible and that will obtain results that you can use in your risk-based program.

9.3.1 □ Objective Personnel Opinion Elicitation Method

People who deal with a plant component on a daily basis, year after year, develop a "feel" for the state of a component and for the changes that have been taking place in that component and its state over time. This "feel" is a ready and knowledgeable information source that you can use to estimate the expected future state of the component.

The objective is to use a proven methodology that will obtain this information in the least threatening way. The information about the state of the equipment or "feel" is intuitive; the result of a subconscious integration that has been taking place throughout the association with the component.

Over the last 20 years, cognitive psychologists who are associated with decision analysis have developed a method that is comprised of a series of questions that are used to tap the integrated information found in the intuition. Sometimes, it is difficult for engineers to accept the value of intuition because of their training and inclination. However, the intuitive information that has been gathering in people who have been continuously associated with equipment is valuable. We should not overlook it, even if it is difficult to objectively obtain it.

The process that follows may appear to be a strange waste of time, however, you are not likely to obtain useful information unless you strictly follow it. All the steps are important. We will give you brief reasons for each step. If you want more information, see (ASME, 1994) and (Hogarth, 1987) for more information and references.

The interview process steps are quite straight-forward. This list briefly discusses each step and the background behind it. The interview subject is referred to as the "operator" and "he," with the understanding that the person could be a mechanic, engineering technician, supervisor, shift engineer or any other position and/or could be female.

If possible, though it is not necessary, try to simultaneously interview two or more people who know the information that you need. This team approach will probably give more accurate information because of the multiple viewpoints that are available. Use a consensus process during the coin stacking. Do not allow voting, because this tends to become adversarial and will inhibit consensus formation. Note that "consensus" means that all interviewees have input and that all interviewees eventually agree. You might need to referee to ensure no subject dominates the decisions. Also, be aware before you begin that consensus building can take a long time, as much as an hour per component, and should not be rushed. You want to seek consensus instead of voting so that you maximize the input from all individuals involved.

Consensus building uses a principal from the total quality methodology. As noted, consensus building seeks maximum input from all the individuals who are involved in the interview process. Further, to ensure ownership, all the individuals need to feel that they have a part in the resulting opinion. If voting is used instead, then the individual or individuals on the losing side of the vote would feel that their opinion was rejected. Also, those on the minority side would have no part of their opinion included in the process outcome.

The interview process proceeds as follows:

1) ***Ask the operator to tell you "his story" about his experience with the component.*** Listening to the operator tell "his story" about what has gone on with the component and tell you about his relationship with it helps him get comfortable with you on this subject and also gets him to focus on the component and its history.
2) ***Ask what the operator's personal exposure would be if component life estimates proved to be in error.*** Knowing what the operator thinks his exposure would be if the life estimate proved wrong provides a basis upon which you may decide whether the operator feels free to express himself. If the interview results later appear to be biased, the operator's perceived exposure may suggest why. For example, an operator who fears death, injury or job loss might bias low; an operator who fears negligence accusations might bias high. The operator's perceived exposure could even help you to decide whether to use him or seek another subject.
3) ***Ask the operator how soon the component could possibly fail.*** Asking about the earliest possible failure date begins to expand his mind.
4) ***Ask the operator "how long the component could possibly last" if it is a single-element component or "when it will no longer be worth fixing" if it is a multiple-element component.*** If it is a multiple-element component then ask him how many elements will fail per year "when it is not worth fixing." Asking about the latest possible failure date further expands his mind and gets him thinking in the other direction.
5) These questions ***will unanchor the operator from any previous life-estimates in which he may have been involved.*** To further unanchor him, use questions that will prompt him to tell stories about why

the component might fail on the earliest date. Getting him to theorize about the component will help him to forget numbers that he might have previously heard, provided, or been given about the expected failure date. Asking for stories about the latest failure date also helps unanchoring. During this step, ask for a couple of stories about each end of the failure date range. Ask him for more stories if he does not appear to be relaxing.

6) ***Get the operator to agree upon some reasonable time increments with which to position the interval between the shortest and longest time to failure.*** This agreement is important. If the time increment is too large, the next step will not have fine enough resolution to effectively reveal the time uncertainty. If the increment is too small, the failure probability consideration at each increment requires too much detailed thinking. Usually, four or five time increments between the earliest and latest failure dates are about right. The only limitation is that the increments must be measured in whole years. Using the previously-agreed upon time increment, prepare a time line that runs from the operator's earliest stated failure date up to his latest failure date.
7) ***Provide the operator with 50 identical coins or washers and ask him to stack them at the points on the time line at which he thinks the component will most likely fail.*** Ask him to stack the coins based on his feeling about when the component will fail, if it is a single element component, or "when it is not worth fixing anymore" if it is a multiple element component. Tell him that he must place at least one coin on each year interval; that otherwise, he can place the coins any way he wishes. (Experience from almost 270 interviews demonstrated that even the most deterministic thinkers provide a fairly broad distribution. None have placed one coin on each interval and all the rest on one interval.)
8) ***Verify that the operator feels comfortable with the stacks, or failure probability distribution, that he has just provided.*** Don't be concerned if the operator is not comfortable with the process; this is not unusual. The important thing is that he is comfortable with the stacks along the time line.

TABLE 9.3.1.1 EXPERT OPINION ELICITATION STEPS

1. Listen to the subject's story about the component.
2. Ask about the subject's exposure in case of an erroneous component life estimate.
3. Ask how soon the component could fail.
4. Ask how long the component could possibly last (single-element component) or when it will not be worth fixing (multiple element component). Then, for a multi-element component, ask how many elements will fail per year when it is not worth fixing.
5. Unanchor the subject from any existing life estimates by asking for stories that illustrate #3 and #4.
6. Get agreement on a reasonable measuring increment.
7. Have the subject stack coins that represent failure likelihood on a time line.
8. Verify comfort with the resulting coin stacks/probability curve.
9. Record the coin stacks.

9) ***Record the number of coins on each time interval for future spreadsheet entry.***

Table 9.3.1.1 lists, in abbreviated form, the steps in the process.

9.3.2 □ Estimating Failure Probability Failure vs. Time Using Interview Data

Next, you will fit a Weibull curve to the interview information. Before you proceed, open Excel and select [Tools] [Add-ins]. A list of the available add-ins will open. If "Analysis ToolPak" and "Analysis ToolPak – VBA" are not checked, click the adjacent boxes to select them and then click [OK]. If the Analysis ToolPak items do not appear in the add-ins list, consult your Excel instructions.

1. If each interval represents a time interval measured in years, take the number of coins on each interval, double it and divide by 100. The resulting numbers are the failure probabilities for each interval. For a multiple element component, multiply the number of elements that will fail per year, when it is not worth fixing, times the probability for each interval and divide by the total number of elements in the component. This puts the probability values on the same basis as they would be for history data or probabilistic models. The minimum interval needs to be a year. This is in order to be compatible with the financial point of view.

TABLE 9.3.2.1 PROBABILITY DATA FOR [FIT_OF_INTERVIEW] COLUMN [A]

The year...	Had this many coins stacked on it...	Doubled...	Divided by 100 and entered in column [A]...
2010	1	2	0.02
2015	5	10	0.1
2020	10	20	0.2
2025	25	50	0.5
2030	9	18	0.18

2a. If you have previously fit a Weibull Distribution to failure history data for this component, call up the previously -saved spreadsheet. (It would have originated as BAYCOM-11.XLS from the Hand-book CD-ROM).

2b. Otherwise, copy BAYCOM-11.XLS from the Handbook CD-ROM to your working directory (use an appropriate name), load it, and open the [Fit_of_ Interview] worksheet tab. The yellow-highlighted areas are the input areas. Place the cursor at cell A5.

3. Enter the individual probability of failure number for each year into column [A]. Table 9.3.2.1 shows how the data that is originally in the spreadsheet was calculated.
4. Enter the year that is associated with each of the interview probabilities into column [C] from cell C5 down.
5. Under the [Fit_of_History] tab, type the Base Year to be used for this data fit into cell G2 if it has not already been entered. The base year is the year of unit startup or the year the component was replaced, which ever occurred later. It is the year that aging began.
6. To fit a regression curve to this interview data, select [Tools] [Fit Interview]. The Weibull-fitted equation alpha and beta will appear in cells E1 and E2. This spreadsheet is now the basis for the failure probability curve that is derived from the interview for the specific component. As before, save this file with a name that indicates to which component it applies.
7. Once this step has been performed for the first component interview, repeat the process for another component for which you will

use interview data. Start with the spreadsheet that contains the history data from Section 7.3 (if applicable); then obtain and process the appropriate interview data.

8. Repeat this process until you have interview-based alpha, beta and base year values for all components for which you need it.

You have now established a failure probability versus time projection curve for each selected component based on interviews.

9.4 □ REMAINING LIFE ASSESSMENT (RLA) MODELING

This section explains how to use engineering data to develop a component failure probability vs. time curve. The engineering data that you will need comes from a kind of analysis that is generally called a remaining life assessment (RLA). If you do not have the expertise that is required for RLA available to you, then you can use this section as a guide to obtaining contract services.

9.4.1 □ Establishing Failure Component Mechanisms

Any remaining life assessment or reliability analysis depends upon the ability to identify the component damage mechanism. This identification usually occurs through a metallurgical failure analysis of the component. Sometimes this analysis can be performed by matching the "macro appearance" of the component failure site with the macro appearance of a previously-analyzed failure or with a macro appearance found among photographs and descriptions that are located in reference handbooks. (ASM, 1975) Component damage mechanism identification is extremely important because the failure mechanism will strongly influence your mathematical model choice. Selecting the wrong failure mechanism can have a major impact on the accuracy of the analysis.

If neither previous failure experience nor a literature search help you to clearly identify the failure mechanism, then you might need to seek metallurgical assistance. Such assistance can be obtained from a competent metallurgical analysis laboratory (ASM, 1975).

If you do not have a previous failure sample, then gather information on the number of service hours, number of start cycles, service temperature and chemical environment. This information can assist in discussions with other companies that have similar equipment or with a consultant about which damage mechanism most likely exists.

Once you have identified a damage mechanism and its associated failure criteria, then you need to obtain an appropriate mathematical model for the mechanism. You can find models for most mechanisms in open-literature references. These references describe the damage mechanism and plot its progression as a function of operation-related parameters. These references have recently been located in volumes that are focused on a specific industry. See, for example, (Viswanathan, 1989)

9.4.2 □ A Spreadsheet-Based Engineering Model for a Component

To illustrate this construction, we have modeled creep rupture in a boiler tube that suffers from a linear-rate build-up of an oxide scale on its inner surface (ID). Such scale insulates the tube and retards heat transfer from combustion to the steam. The operator responds by raising the combustion temperature to maintain the same steam temperature. Continued scale buildup causes a steady temperature increase, which accelerates the creep-rupture damage progression in the tube. The constant steam pressure in the tube is the driving stress.

The deterministic form of this model, with an example, is described in detail in (Viswanathan, 1989). To demonstrate how to develop a probabilistic model for RLA, we expanded that example into a probabilistic model. The spreadsheet that we will use is an MS Excel spreadsheet template that requires the @ Risk add-in to be installed. The spreadsheet may be read without @Risk, but it will not be fully operational.

The example is contained in TUBEMOD.XLS, which is located on the Handbook CD-ROM. The engineering model is built into the spreadsheet, the construction and functioning of which is described in the procedure for its use. @Risk makes a spreadsheet that contains a deterministic engineering model capable of probabilistic analysis. Using @Risk, we will perform a Monte Carlo simulation by assigning random variables (in this case RiskTriang, a triangle distribution function) to specific input variables. This will make the model behave probabilistically when the Monte Carlo simulator is run. The result will be a probability of failure versus time curve that you can input into a Bayesian-like combination analysis (Section 9.5) or use "as-is" in the multiple component ranking based on optimization (Section 10).

Building a deterministic engineering model (base spreadsheet) is beyond the scope of the handbook, however, if you study the example that is provided, you will see the elements that such a model needs to contain and how to modify such a model, once it is built, for probabilistic analysis (Monte Carlo simulation).

The example describes a 2-1/4Cr-1Mo steel boiler tube that operates at a nominal temperature of 1000°F and a hoop stress of 5 ksi. The tube was examined after ten years of service and was found to have an ID (steam side) oxide-scale thickness of 20 mils (0.02 in.).

To analyze this example, proceed as follows:

1. Start Excel by starting @Risk. Use the @Risk icon or (in Windows 95) the Start/Program Menu. If you can't start Excel this way, then consult the @Risk manual.
2. Open the example spreadsheet TUBEMOD.XLS, which is located on the Handbook CD-ROM. This spreadsheet flows from left to right. The areas for input have a yellow background. See Figure 9.4.2.1.
3. The years of interest and the time steps in years are entered in column A.
4. The oxide scale growth, which is assumed to be linear function of time, is calculated in column C.
5. The present oxide thickness is input in cell C2.
6. The Larson-Miller parameter as a function of ID oxide growth is in Column D. This parameter is from Equation 3 in Table 5.6 of (Viswanathan, 1989).
7. In column E, the temperature is calculated from the annual Larson-Miller parameter.
8. In column F, the distribution for tube metal wall temperature is entered as a triangle distribution. A triangle distribution is no doubt simpler than the actual distribution; however, it is easily defined by its "peak" or mode and a variation range. If the analysis proves to depend upon this distribution, we could fit a more accurate "curved"

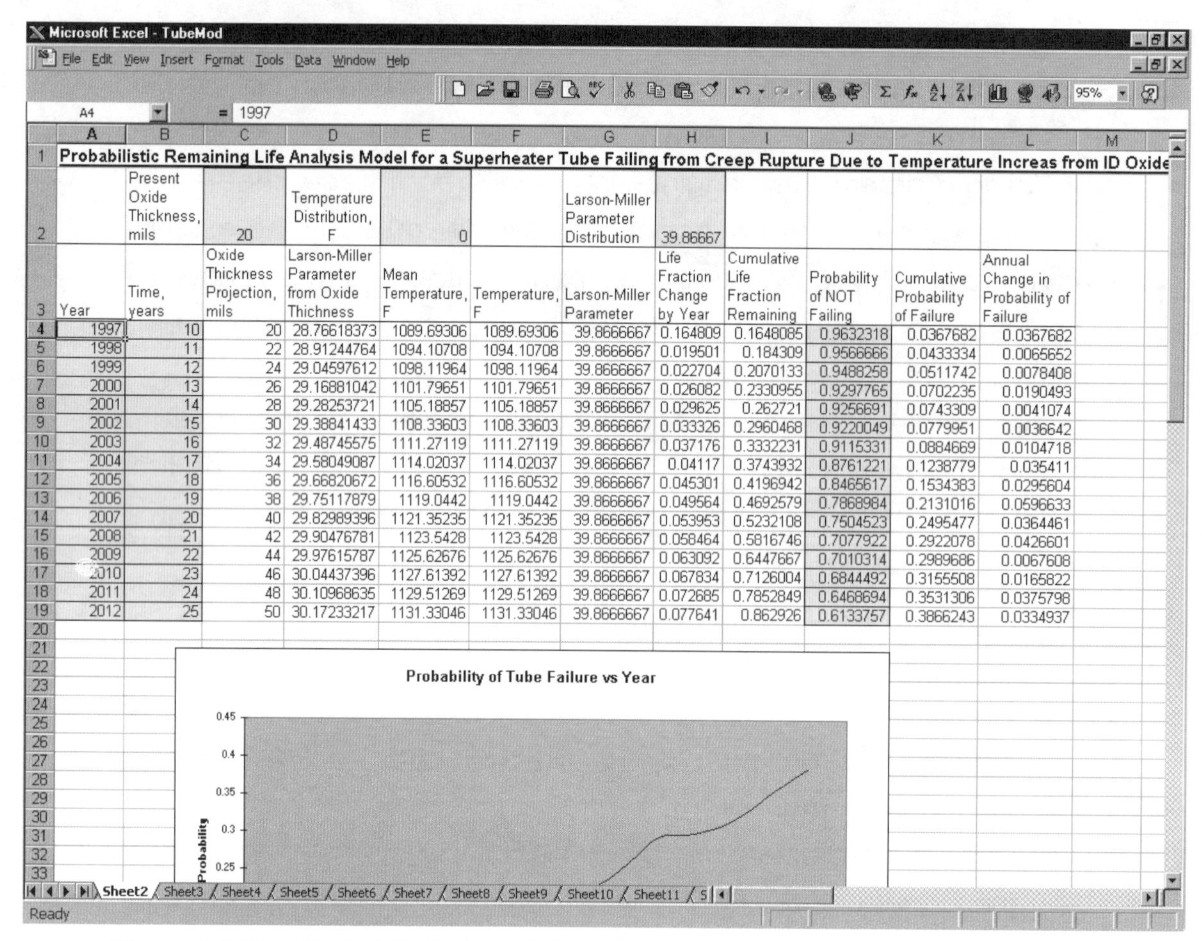

Probabilistic Remaining Life Analysis Model for a Superheater Tube Failing from Creep Rupture Due to Temperature Increas from ID Oxide

	Present Oxide Thickness, mils	20	Temperature Distribution, F	0		Larson-Miller Parameter Distribution	39.86667				
Year	Time, years	Oxide Thickness Projection, mils	Larson-Miller Parameter from Oxide Thichness	Mean Temperature, F	Temperature, F	Larson-Miller Parameter	Life Fraction Change by Year	Cumulative Life Fraction Remaining	Probability of NOT Failing	Cumulative Probability of Failure	Annual Change in Probability of Failure
1997	10	20	28.76618373	1089.69306	1089.69306	39.8666667	0.164809	0.1648085	0.9632318	0.0367682	0.0367682
1998	11	22	28.91244764	1094.10708	1094.10708	39.8666667	0.019501	0.184309	0.9566666	0.0433334	0.0065652
1999	12	24	29.04597612	1098.11964	1098.11964	39.8666667	0.022704	0.2070133	0.9488258	0.0511742	0.0078408
2000	13	26	29.16881042	1101.79651	1101.79651	39.8666667	0.026082	0.2330955	0.9297765	0.0702235	0.0190493
2001	14	28	29.28253721	1105.18857	1105.18857	39.8666667	0.029625	0.262721	0.9256691	0.0743309	0.0041074
2002	15	30	29.38841433	1108.33603	1108.33603	39.8666667	0.033326	0.2960468	0.9220049	0.0779951	0.0036642
2003	16	32	29.48745575	1111.27119	1111.27119	39.8666667	0.037176	0.3332231	0.9115331	0.0884669	0.0104718
2004	17	34	29.58049087	1114.02037	1114.02037	39.8666667	0.04117	0.3743932	0.8761221	0.1238779	0.035411
2005	18	36	29.66820672	1116.60532	1116.60532	39.8666667	0.045301	0.4196942	0.8465617	0.1534383	0.0295604
2006	19	38	29.75117879	1119.0442	1119.0442	39.8666667	0.049564	0.4692579	0.7868984	0.2131016	0.0596633
2007	20	40	29.82989396	1121.35235	1121.35235	39.8666667	0.053953	0.5232108	0.7504523	0.2495477	0.0364461
2008	21	42	29.90476781	1123.5428	1123.5428	39.8666667	0.058464	0.5816746	0.7077922	0.2922078	0.0426601
2009	22	44	29.97615787	1125.62676	1125.62676	39.8666667	0.063092	0.6447667	0.7010314	0.2989686	0.0067608
2010	23	46	30.04437396	1127.61392	1127.61392	39.8666667	0.067834	0.7126004	0.6844492	0.3155508	0.0165822
2011	24	48	30.10968635	1129.51269	1129.51269	39.8666667	0.072685	0.7852849	0.6468694	0.3531306	0.0375798
2012	25	50	30.17233217	1131.33046	1131.33046	39.8666667	0.077641	0.862926	0.6133757	0.3866243	0.0334937

Figure 9.4.2.1 Probabilistic Remaining Life Analysis Model for Boiler Tube Failure in TUBEMOD.XLS

distribution. Note that the distribution treats the temperature that was calculated from the oxide scale thickness as the mode. A plus/minus range is applied in cell E2. This represents the temperature variation across multiple elements or tubes in the boiler.

9. The distribution that represents material creep-rupture behavior is also taken from (Viswanathan, 1989). The spread and mean of the Larson-Miller parameter is taken at 5 ksi. This distribution is input in column H and has a mode of the mean parameter at a stress of 5 ksi and plus/minus range that is determined from the spread of the data.
10. The remainder of column H calculates the ratio of the service hours expended during each year, assuming a 100% service factor, to

the hours to failure calculated for the temperature distribution that was obtained from the oxide thickness. This ratio is called the life fraction. When the ratio of "service hours expended" to "hours to failure" reaches 1.0, it is assumed that the tube will rupture.

11. The cumulative damage fraction as a function of time is shown in column I.
12. The next column is where the probabilistic @Risk analysis will paste in the probability of not failing.
13. The next column is where the cumulative probability of failure is calculated by year.
14. The last column is the change in probability of failure by year.

The probability of failure versus time is obtained by comparing the overlap of the damage propagation distribution and the failure criteria distribution as a function of time. In this particular spreadsheet, the failure criterion is estimated by the Larson-Miller equation calculation of the time distribution to failure based on a 5 ksi stress and the tube metal wall temperature that was estimated for that particular year. The damage propagation is treated as the fraction of the number of service hours expended and is treated as a single value rather than as a distribution. If the estimated service hours expended was considered an uncertainty, then a distribution associated with the 8760 hours in a year would be input in column E.

To review the spreadsheet design:

1. The damage propagation and failure criteria are fed the input they need by the cells on the left side.
2. The failure criteria occupy the cells toward the right.
3. The far right cells estimate the probability of failure by combining these two.
4. To evaluate the damage mechanism progression toward the failure criteria in time, we move down the worksheet. Each row becomes a time step and all the rows together represent the time period of concern.
5. The damage propagation over time is calculated by linking the cells as you move down the rows to form a cumulative damage sum-

mation for each time step. This sum is the cumulative life fraction, column I.

6. The failure probability is based on the time that the damage-to-failure criteria ratio exceeds 1.0 as time progresses row by row. The cumulative combination of the failure criteria and the damage propagation, column I, becomes the output column that @Risk uses to calculate the failure probability versus time.

9.4.3 □ Performing Engineering Failure Probability Versus Time Analysis

1. To perform a probabilistic analysis using TUBEMOD.XLS, ensure that Excel has been loaded with @Risk.
2. Click on the [Display Inputs by Outputs Table] icon. This will display the @Risk page. Note that all the cells that contain distributions are already shown in the input table. The output table is blank.
3. To supply the output cells to @Risk, minimize (hide) the @Risk page. Then drag-select cells L5 through L14. Click on the [Add the selected cells as @RISK outputs] icon and then on the [Display Inputs by Outputs Table] icon. Note that cells I4 through I19 have been added to the output table.
4. Click on the [Sim Settings] button. Select the number of iterations that you want. Usually, this will be the reciprocal of the resolution on the desired order of failure probability times one hundred. For financial problems, 10,000 iterations are usually sufficient. When the component is of safety concern, then 100,000 iterations or more are usually necessary. (For most purposes, we suggest beginning with 1,000 iterations and Monte Carlo sampling. Run the simulation several times, increasing the number of iterations by a factor of ten each time, until the output converges.) Click [OK] when you have set the number of iterations and selected the sampling type. @Risk can stop the analysis when the output converges to within a given percent of change. Because the @Risk convergence can not be set any more closely than 0.5%, it is better to use the iteration limits that are indicated above when you run @Risk.
5. To start the iterations, click on the [Simulate] button. You will see the "Simulating" window in the bottom left of the screen while the

probabilistic analysis is running. When the analysis is complete, the "Results" screen will be displayed.

6. Make active the Simulation Statistics portion of the window by clicking on the title bar. Scroll down the page until "Target#1 (Value)=" appears.
7. Select the first cell next to it, type in "1" and hit [Enter]. Note that a probability is displayed for this target.
8. Now select the cell in which you typed the "1" and over to the right until you reach the last output cell, in this case the cell indicating I19.
9. Then click on [Edit] [Fill To The Right] to copy the target of "1" into the remainder of the Target#1 cells for the rest of the output cells. Note that the probability is automatically calculated in each Target#1 (Probability) cell below it. These probabilities are the cumulative probabilities of not failing. This is because @Risk calculates the probability of a number being below the "Target#1 (Value)=". For the model, the probability of being less than 1.0 is the "Probability of NOT Failing".
10. Now move these probabilities into the "Probability of NOT Failing" column in the spreadsheet. To do this, drag/select the whole row of Target#1 probabilities out to the last output cell or in this case the cell that indicates I19.
11. Select [Edit] [Copy] and click [No] so that only this row of output cell probabilities is copied. Then minimize (hide) the @Risk results screen.
12. Select a remote cell like "M4" in the EXCEL spreadsheet and click [Edit] [Paste] to copy the probabilities into the spreadsheet. Then, to insert these probabilities into column J, select the row of probabilities starting in M4 and [Edit] [Copy]. Then [Paste Special] them into Cell J4, using [Value] [Transpose], and click [OK] to paste these values into the appropriate place.
13. The probabilities of failure are calculated in column K automatically by subtracting the probabilities in column M from 1.0.
14. Note that a plot has been setup below the spreadsheet model to visualize the probability of tube failure versus time by plotting column K vs. column A.

15. The annual changes in the failure probability versus time values are shown in column L. You can copy these values and insert them into spreadsheet BAYCOM11.XLS under the [Fit of History] Tab along with the corresponding year values from column A. Input the year that the component was placed in service or the year that the component was replaced, whichever is later, as the Base Year.

You have now established a failure probability versus time curve for a component based upon an engineering (RLA) analysis. Record the Alpha, Beta and base year values because you will need them in the next step.

If the optimization in Section 10 assigns the component a high enough priority to be inspected, then you can use the spreadsheet model from this section to determine the updated probability of failure versus time curve after the inspection by entering the appropriate component inspection results. The component, with this new curve, can now go again through the risk ranking process (Section 9), and (if it still "makes the cut") the optimization process of Section 10. This is the "update state of knowledge" to which Figure 1.2 refers.

9.4.4 □ A Safety-Related Engineering Failure Probability Versus Time Model

To illustrate this construction, we have modeled creep-fatigue crack growth in a boiler component like steam piping or a header that suffers from a crack, is cyclically loaded, and has a hold time at load during operation. The driving stress can be internal pressure and/or piping system stresses. If piping system stresses are driving the crack, then a supporting analysis, such as a finite element piping system stress analysis, will be needed to calculate those stresses. We have not addressed longitudinal seam weld pipe cracks in this example. The deterministic form of the elements of this model is described in (Viswanathan, 1989). To demonstrate a safety-related example of a probabilistic model for RLA, we have provided a deterministic model and then expanded it into a probabilistic model.

The example describes a 2-1/4Cr-1Mo steel pipe that operates at a nominal temperature of 1005° F and an internal pressure of 900 psi. We have assumed a radial-axial ID connected crack that is not located in a seam weld. The pipe was examined after ten years of service.

To analyze this example, proceed as follows:

1. Start Excel by starting @Risk. Use the @Risk icon the Start/Program Menu. If you can't start Excel this way, then consult the @Risk manual.
2. Open the example spreadsheet PIPESAFE.XLS, which is located on the Handbook CD-ROM. This spreadsheet flows from left to right. The areas for input have a yellow background.
3. The years of interest and the time steps (in years) are entered in column A.
4. The service hours projected per year are entered in column B.
5. The starts per year for the unit are entered in column C.
6. Enter the pipe I.D., internal pipe pressure and its distribution, the wall thickness, the start and stop stress ratio (minimum stress/maximum stress), the service temperature and its distribution, and the initial crack depth and the distribution on initial crack depth at the top of columns C, E, G, I and K.
7. In column D, the average hold time at load is calculated, and the variation in stresses are calculated in column E.
8. In column F, the Maximum stress intensity is calculated.
9. Column G contains Cstar (C^*), which is the steady state driving force for creep crack growth.
10. The next column, column H, calculates the fatigue crack growth per start.
11. Column I is the transient creep crack growth rate per start considering the creep crack growth from loading and operation during and after a start.
12. The next columns, columns J and K, are the total crack growth per year and the total annual cumulative crack depth, in inches.
13. Column L calculates the ratio of the remaining net section stress as a result of through wall crack growth to the material flow stress. When this ratio exceeds 1.0, failure would occur with the crack penetrating through the wall and resulting in leak.
14. Column M is the probability of not failing, which is pasted in from the @Risk analysis.
15. Column N is the resulting cumulative probability of failure by year.

16. The top of column M is the steady state or secondary creep growth coefficient for the strain rate relationship and column O is the steady state or secondary creep crack exponent, both of which appear in the Cstar relationship.
17. Cell O3 is the transient creep crack exponent. At the top of column Q is the transient creep crack coefficient for the C(t) creep crack growth relationship and its distribution. The C(t) relationship is in the CRACK_TS macro.
18. The top of column S is the fatigue crack growth coefficient and its distribution and the top of column U is the fatigue crack growth exponent, both of which appear in the Paris equation. Also in column U is the failure criterion of material flow stress.

The probability of failure versus time is obtained by comparing the overlap of the damage propagation distribution and the failure criteria distribution as a function of time. In this particular spreadsheet, the failure criterion is the flow stress in the component material. The damage propagation is treated as the total crack depth, which is a function of time and is obtained by adding the fatigue crack growth and creep crack growth components . The probability of failure is determined from the net section stress to flow stress ratio exceeding 1.0.

To review the spreadsheet design:

1. The damage propagation calculation is fed the input that it needs by the cells on the left side.
2. The failure criterion occupies the cell on the far top right.
3. The far right cells estimate the probability of failure by combining these two.
4. To evaluate the damage mechanism progression toward the failure criteria in time, we move down the worksheet. Each row becomes a time step and all the rows together represent the time period of concern.
5. The damage propagation over time is calculated by linking the cells as you move down the rows to form a cumulative crack depth for each time step. Note that the annual crack depth is the total of the annual fatigue and creep components. For this particular case, the creep crack growth rate component needs to be integrated over

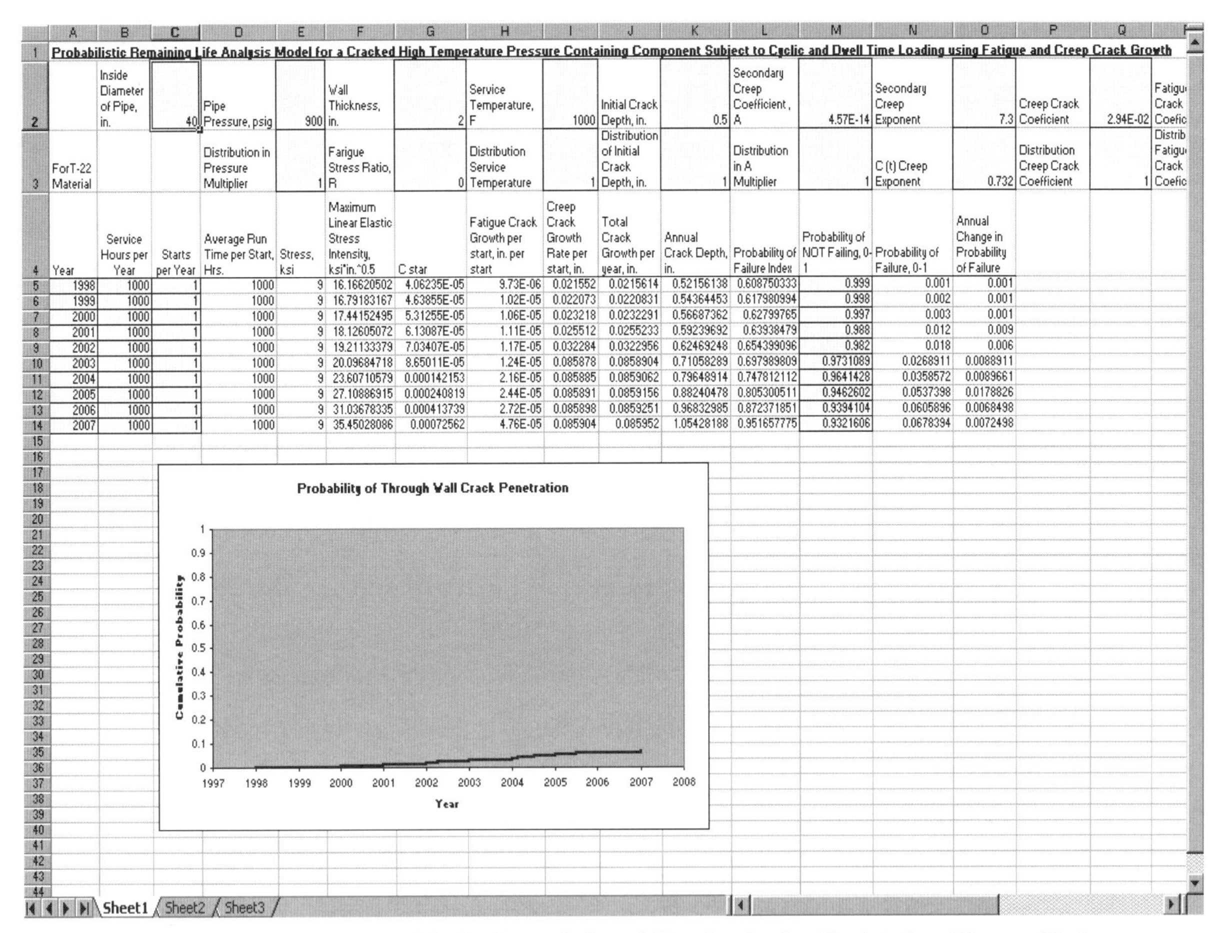

	A	B	C	D	E	F	G	H	I	J	K	L	M	N	O	P	Q	R
1	Probabilistic Remaining Life Analysis Model for a Cracked High Temperature Pressure Containing Component Subject to Cyclic and Dwell Time Loading using Fatigue and Creep Crack Growth																	
2		Inside Diameter of Pipe, in.	40	Pipe Pressure, psig	900	Wall Thickness, in.	2	Service Temperature, F	1000	Initial Crack Depth, in.	0.5	Secondary Creep Coefficient , A	4.57E-14	Secondary Creep Exponent	7.3	Creep Crack Coeficient	2.94E-02	Fatigu Crack Coefic
3	ForT-22 Material			Distribution in Pressure Multiplier	1	Farigue Stress Ratio, R	0	Distribution Service Temperature	1	Distribution of Initial Crack Depth, in.	1	Distribution in A Multiplier	1	C (t) Creep Exponent	0.732	Distribution Creep Crack Coefficient	1	Distrib Fatigu Crack Coefic
4	Year	Service Hours per Year	Starts per Year	Average Run Time per Start, Hrs.	Stress, ksi	Maximum Linear Elastic Stress Intensity, ksi*in.^0.5	C star	Fatigue Crack Growth per start, in. per start	Creep Crack Growth Rate per start, in.	Total Crack Growth per year, in.	Annual Crack Depth, in.	Probability of Failure Index	Probability of NOT Failing, 0-1	Probability of Failure, 0-1	Annual Change in Probability of Failure			
5	1998	1000	1	1000	9	16.16620502	4.06235E-05	9.73E-06	0.021552	0.0215614	0.52156138	0.608750333	0.999	0.001	0.001			
6	1999	1000	1	1000	9	16.79183167	4.63855E-05	1.02E-05	0.022073	0.0220831	0.54364453	0.617980994	0.998	0.002	0.001			
7	2000	1000	1	1000	9	17.44152495	5.31255E-05	1.06E-05	0.023218	0.0232291	0.56687362	0.62799765	0.997	0.003	0.001			
8	2001	1000	1	1000	9	18.12605072	6.13087E-05	1.11E-05	0.025512	0.0255233	0.59239692	0.63938479	0.988	0.012	0.009			
9	2002	1000	1	1000	9	19.21133379	7.03407E-05	1.17E-05	0.032284	0.0322956	0.62469248	0.654399096	0.982	0.018	0.006			
10	2003	1000	1	1000	9	20.09684718	8.65011E-05	1.24E-05	0.085878	0.0858904	0.71058289	0.697989809	0.9731089	0.0268911	0.0088911			
11	2004	1000	1	1000	9	23.60710579	0.000142153	2.16E-05	0.085885	0.0859062	0.79648914	0.747812112	0.9641428	0.0358572	0.0089661			
12	2005	1000	1	1000	9	27.10886915	0.000240819	2.44E-05	0.085891	0.0859156	0.88240478	0.805300511	0.9462602	0.0537398	0.0178826			
13	2006	1000	1	1000	9	31.03678335	0.000413739	2.72E-05	0.085898	0.0859251	0.96832985	0.872371851	0.9394104	0.0605896	0.0068498			
14	2007	1000	1	1000	9	35.45028086	0.00072562	4.76E-05	0.085904	0.085952	1.05428188	0.951657775	0.9321606	0.0678394	0.0072498			

Figure 9.4.4.1 Probabilistic Remaining Life Analysis Model for Steam Piping Failure in PIPESAFE.XLS

the dwell time. A macro named CRACK_TS, which can be viewed in the Visual Basic editor, performs this task.

6. The failure probability is based on the time that the net section stress-to-flow stress ratio exceeds 1.0 as time progresses row by row. The cumulative combination of the failure criteria and the damage propagation, column L, becomes the output column that @Risk uses to calculate the failure probability versus time.

9.4.5 □ Performing Safety-Related Engineering Failure Probability Versus Time Analysis

1. To perform a probabilistic analysis using PIPESAFE.XLS, ensure that Excel has been loaded with @Risk.

2. Click on the [Display Inputs by Outputs Table] icon. This will display the @Risk page. Note that all the cells that contain distributions are already shown in the input table. The output table is blank.
3. To supply the output cells to @Risk, minimize (hide) the @Risk page. Then drag-select cells L5 through L14. Click on the [Add the selected cells as @RISK outputs] icon and then on the [Display Inputs by Outputs Table] icon. Note that cells L5 through L14 have been added to the output table.
4. Click on the [Sim Settings] button. Select the number of iterations that you want. For financial problems, 10,000 iterations are usually sufficient. When safety is a concern, then 100,000 iterations or more are usually necessary. (We suggest beginning with 1,000 iterations and Monte Carlo sampling. Run the simulation several times, increasing the number of iterations by a factor of ten each time, until the output converges.) Click [Macro] tab when you have set the number of iterations.
5. Click on [ExecuteMacro?] check box. Type "crack_ts" in the Macro name. Click on [After sampling/worksheet recalc] radio button. Click [OK].
6. To start the iterations, click on the [Simulate] button. You will see the "Simulating" window in the bottom left of the screen while the probabilistic analysis is running. When the analysis is complete, the "Results" screen will be displayed.
7. Make active the Simulation Statistics portion of the window by clicking on the title bar. Scroll down the page until "Target#1 (Value)=" appears.
8. Select the first cell next to it, type in "1" and hit ⟨Enter⟩. Note that a probability is displayed for this target.
9. Now select the first cell in which you typed in the "1" and over to the right of the cell where you entered the "1" until you reach the last output cell, in this case the cell indicating L14.
10. Then click on [Edit] [Fill To The Right] to copy the target of "1" into the remainder of the Target#1 cells for the rest of the output cells. Note that the probability is automatically calculated in each Target#1 (Probability) cell below it. These probabilities are the cumulative probabilities of not failing. This is because @Risk calculates the

probability of a number being below the "Target#1 (Value)=". For the model, the probability of being less than 1.0 is the "Probability of NOT Failing".

11. Now move these probabilities into the "Probability of NOT Failing" column in the spreadsheet. To do this, drag/select the whole row of Target#1 probabilities out to the last output cell or in this case the cell that indicates L14.
12. Select [Edit] [Copy] and click [No] so that only this row of output cell probabilities is copied. Then minimize (hide) the @Risk results screen.
13. Select a remote cell like "R5" in the EXCEL spreadsheet and click [Edit] [Paste] to copy the probabilities into the spreadsheet. Then, to insert these probabilities into column M, select the row of probabilities starting in R5 and [Edit] [Copy]. Then [Paste Special] them into Cell M5, using [Value] [Transpose], and click [OK] to paste these values into the appropriate place.
14. The cumulative probabilities of failure are calculated in column N automatically by subtracting the probabilities in column M from 1.0.
15. Note that a plot has been setup below the spreadsheet model to help you to visualize the cumulative probability of crack wall penetration versus time by plotting column N vs. column A.
16. The annual change in failure probability versus time values are shown in column O. You can copy these values and insert them into spreadsheet BAYCOM11.XLS under the [Fit of History] Tab along with the corresponding year values from column A. Input the year that the component was placed in service or the year that the component was replaced, whichever is later, as the Base Year.
17. From this spreadsheet, select [Tools] [Fit History]. This macro will fit the remaining life assessment (RLA) data that you have entered with a three-parameter Weibull distribution equation.
18. Note the Alpha, Beta and Base Year parameter values that resulted for this component. You will use these parameters in subsequent analysis to describe the Engineering Model projection for this component.

You have now established a failure probability versus time curve for a safety related component based upon an engineering (RLA) analysis. Record the Alpha, Beta and base year values because you will need them in the next step.

9.5 □ THE BAYESIAN-LIKE COMBINATION PROCESS CONCEPT

We need an objective and consistent way to combine failure-probability-versus-time information that is derived from different sources. The Bayesian-like combination process, in simple terms, modifies old information to account for newly-obtained knowledge or information. Because the information with which we are concerned is in the form of probability curves, we need to use a combining method that obeys the probability theory axioms. The Bayesian-like combination process does this. We call it "Bayesian-like" because, although it is not strictly the Bayesian process, it is derived from the Bayesian approach.

There is a subtlety about which you should be aware when you apply and interpret the Bayesian combination analysis. The distribution in which you have prior knowledge or belief is usually called the prior distribution. In this approach, it must be a full 0-to-1 probability distribution in the time frame over which you are combining data. However, the "new" distribution in the combination is not necessarily a 0-to-1 probability distribution in the same time frame. This second distribution is called the updating distribution. It is the new knowledge that we have gained about the component behavior during the time frame of concern. It occupies this position because we have little control over the time frame during which it will change from 0 to 1. The resulting combined distribution can be called the posterior distribution. Because the prior distribution must have a 0-to-1 probability distribution in the time frame of concern, it will usually have a higher weight or impact on the resulting combined 0 to 1 probability distribution in this time frame than will the updating distribution. This is because the distribution with the highest rate of change in a given year will dominate the product of the two distributions.

The Bayesian-like combination process for two failure probability versus time curves combines the two curves (prior and updating probability distributions) in the same time frame and then normalizes them to reflect the new updated failure probability versus time within the same time frame. In probability theory, the joint probability of coexistence of two events, like two failure probabilities, is the product of the two probabilities at a given time. To obtain the new probability distribution, confined to the time frame of concern, and because the prior distribution was 0 to 1 during this time frame, we divide each year's joint probability by the sum of all the joint probabilities for the whole time frame. This provides the complete updated failure probability versus time curve, or posterior probability distribution, for the whole time frame.

This process is illustrated with the following table:

Year	Prior Distribution	Updating Distribution	Joint Probabilities: row-by-row product of the Prior and Likelihood Probabilities	Posterior Distribution: normalized joint probabilities, row by row
1998	0.1	0.01	0.1 × 0.01 = 0.001	0.001/0.05 = 0.02
1999	0.2	0.03	0.2 × 0.03 = 0.006	0.006/0.05 = 0.12
2000	0.4	0.05	0.4 × 0.05 = 0.02	0.02/0.05 = 0.4
2001	0.2	0.07	0.2 × 0.07 = 0.014	0.014/0.05 = 0.28
2002	0.1	0.09	0.1 × 0.09 = 0.009	0.009/0.05 = 0.18
			Sum of Joint Probabilities 0.05	

Note how the spread of the prior distribution is symmetrical about the year 2000, while the updating distribution is constantly increasing from 1998 to 2002, even though it is at low values for all the years. The result of the normalization process of the joint distribution or Bayesian-like combination is the posterior distribution with its density shifted toward 2001 and 2002.

Three data sources are described in this handbook: failure history, engineering analysis (RLA) and plant personnel (expert) opinions. You can use the procedures in Section 9.5.1 and 9.5.2 to combine data from any two or from all three sources. Proceed as follows:

Case 1: Combine failure history and expert opinion

In this analysis, you will combine plant personnel belief as expressed by interview information with failure history information. When you combine failure history and interview information, you should treat the interview information as the prior distribution and let its distribution control the whole time frame of the update.

We take the time frame of the interview information and the failure history information in this time frame and perform the one-step Bayesian-like combination analysis. This produces a new distribution that represents the plant personnel belief of the failure probability combined with the failure history.

Case 2: Combine failure history and RLA results

In this analysis, you will combine the engineering belief as expressed by the RLA model with the failure history information. When you combine failure history and RLA model results, you should treat the RLA model results as the prior distribution and let its distribution control the whole time frame of the update.

We take the time frame of the RLA results and the failure history information in this time frame and perform a one-step Bayesian-like combination analysis. This produces a new distribution that represents the engineering belief in the failure probability combined with the failure history.

Case 3: Combine expert opinion and RLA results

In this analysis, you will combine plant personnel belief as expressed by interview information with RLA results. When you combine interview information and RLA results, you should treat the interview information as the prior distribution and let its distribution control the whole the time frame of the update.

We take the interview results time frame and the RLA model results in this time frame and perform the one-step Bayesian-like combination analysis. This produces a new distribution that represents the plant personnel belief in the failure probability updated by the RLA results.

Case 4: Combine failure history, expert opinion and RLA results

There are three sources of information and we want to consider all three when we estimate the failure probability versus time curve.

We use the personnel opinion failure probability versus time curve as the prior distribution in a first combination with failure history failure probability versus time curve being the updating distribution. Then, in a second combination, we use the posterior from the previous combination as the prior distribution and the RLA results as the updating distribution.

9.5.1 □ Spreadsheet for Combining Two Data Sources

Before you proceed, if you have not done so already, open Excel and select [Tools] [Add-ins]. A list of the available add-ins will open. If "Analysis ToolPak" and "Analysis ToolPak – VBA" are not checked, click the adjacent boxes to select them and then click [OK]. If the Analysis ToolPak items do not appear in the add-ins list, consult your Excel instructions.

1a. Copy the spreadsheet BAYCOM11.XLS from the Handbook CD-ROM to your working directory. Use a suitable name. Open the spreadsheet and load the failure history data, as described in Section 4, and the failure interview data, as described in section 9.3.2. or
1b. Call up the spreadsheet that you have already created (based upon BAYCOM11.XLS) for a component that contains previously-analyzed failure history and interview data.
2. Go to the tab [Bayesian Combination].
3. The likelihood distribution information from the failure history is automatically transferred from the [Fit_of_History] tab when you run the Weibull fit macro with [Tools] [Fit History]. Likewise, the "prior" distribution information from the interview data is automatically transferred from the [Fit_of_Interview] tab when the Weibull fit macro is run with [Tools] [Fit Interview].
4. Under tab [Bayesian Combination], run the Bayesian-like combination analysis macro by selecting [Tools] [Bayesian Combine], and see the resulting posterior distribution plot.
5. Next, select [Tools] [Fit Bayesian] to perform a three-parameter Weibull fit to the posterior distribution data and display the resulting plot. The alpha, beta and base year parameters for the three-parameter Weibull distribution are shown in cells E7, G7, and K5 of this worksheet.

6. Record the alpha, beta and base year parameters for this component.

9.5.2 □ Spreadsheet for Combining a Third Data Source

Before you proceed, if you have not done so already, open Excel and select [Tools] [Add-ins]. A list of the available add-ins will open. If "Analysis ToolPak" and "Analysis ToolPak – VBA" are not checked, click the adjacent boxes to select them and then click [OK]. If the Analysis ToolPak items do not appear in the add-ins list, consult your Excel instructions.

1. Load a new copy of BAYCOM12.XLS. Copy it to your working directory under a suitable name.
2. Go to tab [Prior] and copy your prior distribution (the previously-combined interview and failure history) into column G from row 5 to but not including the "0" probability. Note that the probability will go from near 0 to 1 in the time frame you are analyzing. Copy the associated year numbers from Column C. Put the base year into cell G2 of tab [Updating].
3. Enter your third probability (the RLA data) into column C starting at cell C5 under tab [Updating]. Make sure that there are entries for each year through the last year for the prior distribution. Enter the associated year numbers in column A.
4. Run the macros that will perform three-parameter Weibull fits to the data in both distributions. Select [Tools] [Fit Prior] under the [Prior] tab and select [Tools] [Fit Updating] under the [Updating] tab. The alpha, beta and base year parameters for each Weibull distribution appear in cells E1, E2 and G2 the respective worksheet.
5. Under tab [Bayesian Combination], run the Bayesian-like combination analysis macro by selecting [Tools] [Bayesian Combine] and see the resulting posterior distribution plot.
6. Next, select [Tools] [Fit Bayesian] from the menu to perform a three-parameter Weibull fit to the posterior distribution data and see the resulting plot. The alpha, beta and base year parameters for the three-parameter Weibull distribution are shown in cells E7, G7 and K5.
7. Record the alpha, beta and base year parameters for this component.

9.6 □ DETERMINING CONSEQUENCES

Before you translate the failure probability into expected forced outage consequences, you need to understand the component operational function and how the overall system depends on the component operation. If a component is made up of a lot of elements and the loss of production is a function of the element failure frequency, then estimation of the consequence is fairly straight forward. The total expected failure consequence is merely:

$$C_f = (P_f)ntpc$$

Where C_f = total expected failure consequences

P_f = cumulative failure probability

n = number of elements

t = lost production time per failure, hr.

p = production lost per hour with the failure of an element

c = cost per unit of production loss

The failure probability should be for the time period that has been chosen as the analysis period, which will probably be the same as the financial analysis time window.

If the number of elements is small or one, then the consequence is the cumulative failure probability times the consequence of the monolithic or near-monolithic component failing. This usually has to be estimated because of the lack of failure experience with these components. In this case, the consequence is usually based upon the estimated number of production shutdown hours that component failure would cause.

For the probability axis on the risk plot, it is best to use the cumulative failure probability for the failure occurrences axis, if the elements, like boiler tubes, would cause a unit shutdown. This is in order to use a common scale for occurrences for multiple and monolithic element components.

9.7 □ TRANSLATING FAILURE PROBABILITY INTO FORCED OUTAGE CONSEQUENCES AND OCCURRENCES

1. Load example spreadsheet CONSEQ.XLS from the Handbook CD-ROM. Save it to your working directory under a suitable name. See Figure 9.7.1.
2. In cell B3, enter the present year for this analysis.
3. Enter the component name in cell D3, so that you can keep track of your inputs later.
4. Insert the Alpha, Beta and Base Year from the previous analysis for this component.
5. In cell B5, enter the number of independent component elements
6. In cell D5, enter the number of production hours that are lost when the component fails until the unit is back in production.
7. In cell F5, insert the production loss per unit time. This is a known value; for power production, it will be MW. If the failure of this component will cause a complete loss of production, enter the maximum dependable capacity for the unit. If failure of the component would not cause complete loss of production, then estimate what the loss amount would be for the failure of this component and enter the estimate. Or, go to Section 8 and use fault tree/event tree analysis to calculate the loss amount.
8. In cell H5, enter the cost per unit of production loss. This is usually a known value and is usually measured in dollars.
9. Note in cell J6 the total consequences in lost production for this component and in cell I6 the total failure probability for the analysis period that we are using. These two numbers will now be used in the risk ranking.
10. Repeat this process for all the components of concern.

9.8 □ RISK ASSESSMENT

In this section, you will perform a quantitative risk assessment and screening. You can use either a risk-ranking table or a risk plot (or both!).

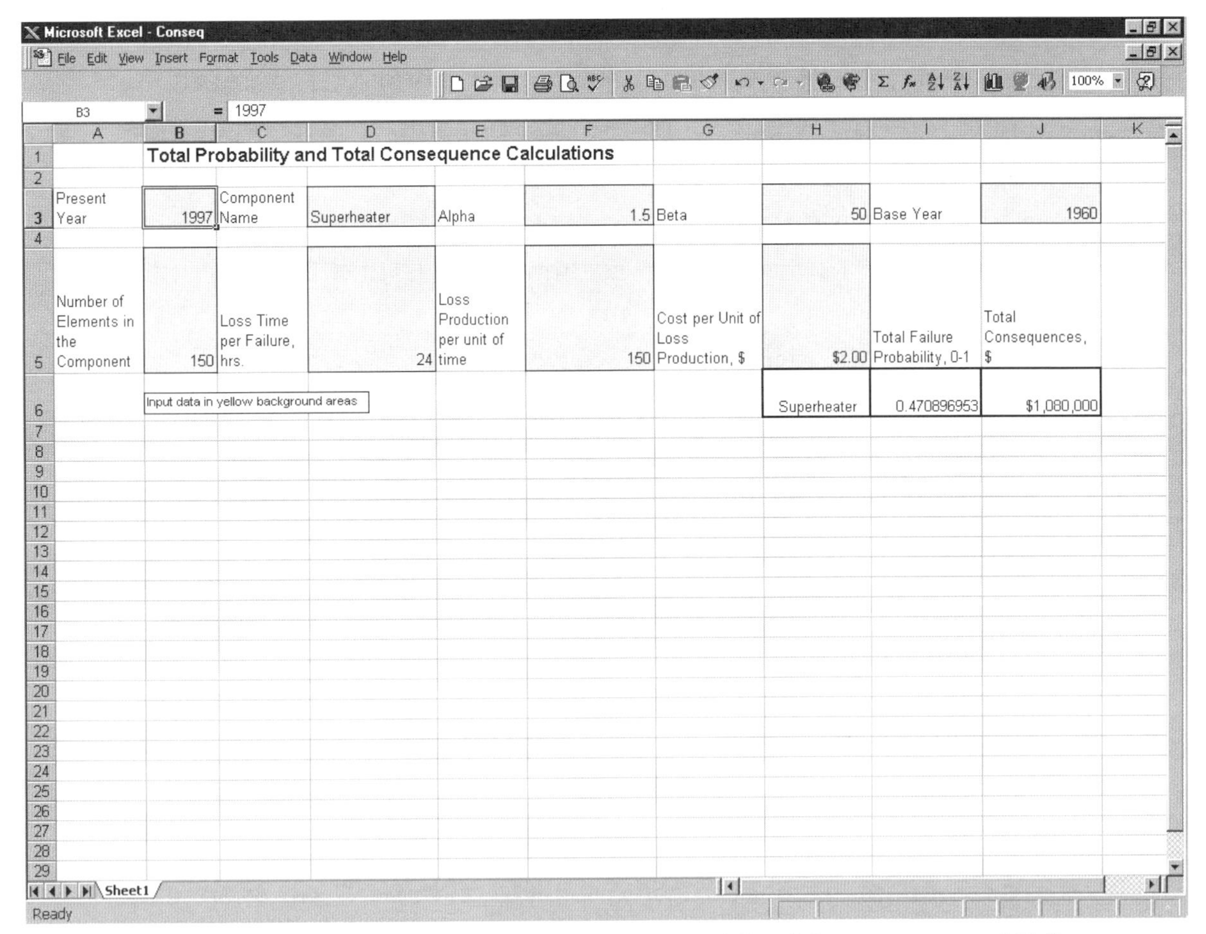

Figure 9.7.1 Calculation of Total Probability and Total Consequence of Failure in CONSEQ.XLS

9.8.1 □ Assessment by Risk Ranking Table

Now that you have selected the components of concern and arranged them in the consequence spreadsheet format, it is time to rank them. Note that when the macro was building the spreadsheet, as each line was inserted, the failure probabilities were multiplied by the total consequences, providing the risk that you see in column C. First, we will risk-rank by sorting.

1. In the same Excel session that is running the filled spreadsheet that started as CONSEQ.XLS, load the RISKPLT$.XLS spreadsheet from the CD-ROM. Use [File] [Save as] to copy it to your working directory with a suitable name.
2. [Edit][Copy] from row 6 columns H through J of the consequence

spreadsheet for each component or cause code. [Edit] [Paste Special] [Value] on subsequent lines of the RISKPLT$ spreadsheet until you have included all the components on which you have performed the consequence analysis. Note that the Risk for each component is automatically calculated in column D.

3. Drag-select all the rows of columns A, B, C and D that contain data.
4. Click on [DATA] [SORT].
5. Click on the primary sort field (Risk, column D), and select [Descending order].
6. Click on [OK] and the sort will run.
7. Save this file. It is a sorted list of risk ranked components of concern.
8. Inspect the table that you have created. There are probably a number of components (cause codes) at the bottom of the ranked table that have much lower risk levels than those that proceed them in the list. You will want to delete these components and focus your attention on the most important items.
9. If, after you drop obviously lower-risk items, there are still too many components (cause codes) to analyze on the first pass, then select the top 20% of the components for the first pass and delete the rest. The 20% criterion, which is from Pareto was introduced in Section 6. Apply this thumb rule by counting the components (cause codes) and deleting columns A through C for all except the "top 20%."
10. Save this file under a different name. It is a sorted and culled list of risk ranked components of concern

Do not be concerned about the 80% of components that you are dropping. These components will be reexamined after the inspection program has been prioritized and developed for the selected "first pass" components.

9.8.2 □ Assessment by Risk Plot

1. Load the file, which was based on RISKPLT$.XLS, that you saved in Section 9.8.1 step 7.

2. Double click on the risk plot, which is located just above and to the right of the risk table.
3. This plot has the cumulative probability of failure on the vertical axis and the total dollar loss consequences on the horizontal axis. In order to place the correct labels by the data points in column A, select the chart by clicking on it and then click on the button titled [LABEL PLOT POINTS].
4. A log-log plot also appears below the previously-described plot. In order to place the correct labels by the data points in column A, select the chart by clicking on it and then click on the button titled [LABEL PLOT POINTS]. The objective now is to set on a risk plot a line that will segregate the components that need further study this round. You can place the line anywhere on the risk plot by selecting the plot, clicking on the line and moving the ends of the line with the left mouse button.

   If, after you have placed the line, the product of a point on the vertical axis with its corresponding point on the horizontal axis is equal to the product of a point on the horizontal axis and its corresponding point on the vertical axis, then you have a line of constant risk when these end points are joined with the line. The easiest way to place the line is to choose, by engineering judgment, a number of occurrences and/or total consequence that seems excessive at this pass.

The line that you have just located is the minimum risk line. The components or cause codes that are to the right of this line identify the components that you will carry further in inspection prioritization and program development. Those to the left of the line you can ignore until you are choosing a set of components for the next pass.

Record the Base Year and the Weibull alpha and beta and the consequence cost information for each component that will be carried further in the analysis into Sections 10 or 11.

NOTES:

SECTION 10

To perform this procedure, you will need:

- *Your component ranking list, sorted by risk-significance, with screening criteria selected and "first cut" components identified. See Section 7.2 and/or 9.8.*
- *Weibull parameters for the identified components' failure probability versus time curves, see Section 7.3, or "combined" Weibull parameters, see Section 9.5.*
- *For components for which the relationship between component failure and mission failure is complex, a derived mission failure probability vs. time curve. See Section 8.*
- *Repair or replacement costs for the components of interest.*

To tailor this procedure to your facility, you will need:

- *The value measure that your facility uses for financial decision making. The example spreadsheet assumes maximum net present value (NPV). You can modify the NPV assumptions or you can modify the spreadsheet to calculate a different value measure.*
- *Your facility safety limit, if any of the components that you are considering are safety related. The example spreadsheet assigns a value of 2×10^{-4}. You can change this value, or you can modify the spreadsheet to use a different safety-constraining strategy.*
- *Projected demand for the unit(s) that contain the selected components.*
- *Lost production cost for the unit(s) that contain the selected components.*

This procedure will:

- *Properly rank the selected components for inspection.*
- *Determine when to inspect and/or replace the selected components.*
- *Calculate the NPV for the plan.*

The next step will, in general, be inspection program development. You could, however, find that none of the components of interest require attention soon enough to concern you. In this case, you would pick another "problem" and start the process over.

MULTIPLE COMPONENT RANKING BASED UPON OPTIMIZATION 10

This (rather horribly titled) section is the heart of the risk-based process as it is described in this handbook. All the previous procedures had no higher purpose then to deliver a set of failure probability vs. time curves for the most risk-significant components to this section. As you will learn, inspections are not themselves important. Their main purpose is to validate the conclusions you will reach in this section.

Obviously cleaner candidates for the title of this section are "Financially optimized component risk-ranking and timing" or "maintenance optimization." Those don't work, because safety is a critical piece, so we would have to add somewhere "safety (and environmentally, and maybe otherwise) constrained." There is actually one error in the title, as focused as it might seem: you can "optimize" a single component. The timing calculation that is performed in this section adds value to the risk-ranking.

In this section, you will calculate the best time to take action on one or more components. The best time will be the best time because it is the time that maximizes NPV while it meets the safety and other selected constraints. You need to work this step before you start worrying about inspection programs because inspection program development is a resource-intensive process that you do not want to undertake until you know what it will do for you.

The "action" upon which this section is based is component replacement. We use replacement cost instead of inspection cost because inspection cost is so small compared to outage costs that an optimizing calculation that is based upon inspection cost would say, "inspect everything now!" The strategy that we suggest:

1. Optimize (with safety and other constraints) based upon replacement
2. Inspect (using an optimized strategy, of course) before the projected component replacement date
3. Compare the actual component conditions that are found with the projected conditions
4. Replace the component or calculate a new optimized replacement date.

10.1 □ NEED FOR OPTIMIZATION

This section introduces the "why" and basic principles of optimization. It also discusses ways to find various financial information that you will need and, if possible, what to use for defaults if you cannot obtain the information. Finally, it continues the discussion of financial methods that began in the introduction.

10.1.1 □ Knowing What to Inspect and When to Inspect It

You have a list of components that you reasonably expect are the most risk-significant. However, the only consequence that you have considered so far is an estimated production (MWH) loss. This section tells you how to use more sophisticated, finance-based analysis techniques to prioritize the components in a way that ensures you do not exceed safety or budgetary constraints. The process will also:

- Determine the most financially sound inspection year for each component
- Provide systematic and easily reviewable documentation that avoids acceptance of unjustified assumptions
- Help detect overstated or otherwise erroneous inputs

This optimization process for multiple components provides the greatest value when it is used to plan the major repair or replacement of components that are amenable to predictive maintenance techniques, high in replacement cost and are in the aging-dominated part of their life cycle.

10.1.2 □ Determining the Value Measure, Safety Limits and Failure Consequences

This procedure prioritizes components by calculating the component replacement order and timing that will produce the highest net present value (NPV) that can be achieved with the failure probability below the selected safety limit and without spending more money than the budget allows during any year. If you accept the value measure and the safety limit that are given in the introduction, skip to Section 10.2. Otherwise, you will need to gather some information.

The first piece of information you will need is the value measure that your corporation uses to make financial investment decisions. To obtain this information, find out who does financial analysis for company investment decisions. This person is usually located in the financial or investment part of the company and serves as a financial analytical advisor to senior management.

When you have identified that person, explain that you are developing an inspection program that will maximize value. Discuss the value measure(s) that the company uses for decision making. If you do not obtain a clear answer, you might suggest NPV. NPV is a commonly accepted value measure for financial decision making. It is used in many corporate financial texts. (See Brealey, 1991, and Brigham, 1991.) In fact, we suggest that you use NPV unless you encounter strong resistance.

You should discuss the approach to value measure that the example uses. If that approach is not acceptable, you should ask to be guided through the corporate NPV calculation process. You might ask the corporate financial analyst to help you to modify the financial portions of the example spreadsheets, particularly if you need to use a value measure other than NPV.

If you adjust the spreadsheets, confirm with your local management that the selected value measure and the way that you are calculating it are suitable.

Next, you will need the facility safety limit. The calculations that the examples use to prioritize inspection programs will allow you to determine whether the expected failure probability exceeds this limit.

The safety limit is likely to be more difficult to obtain than was the value measure, because many people do not understand risk. "Zero tolerance" (for safety-related risk) may be a good principle for public-relations purposes, but it is not practical for internal decision making. Any activity, like working in an industrial facility or even driving an automobile, involves some probability of injury or death. It is important to choose as the safety limit a realistic probability value.

You may need to give management examples of every-day fatality probability values. For example, since the imposition of the 55 mph speed limit (about 1980), an individual's probability of being killed each year in a car has been about 2×10^{-4}. Point out that setting the facility safety limit at this level implies that the inspection program will not impose a risk burden any worse than that which is already accepted by persons who choose to drive to work. Another supporting statistic is that a transportation or utility worker has (since about 1950) had about the same (2×10^{-4}) probability of dying on the job during each year worked. This information may facilitate an agreement. At least it will start a dialog with management from which may emerge an agreed limiting value.

The actual safety risk that is implied in Handbook-based calculations is much less than the failure probability safety limit that you select, because most failures will not directly cause a fatality. The Handbook calculations, however, conservatively assume that this is so. If you cannot accept the assumption that failure and fatality (or injury) probabilities are the same, you will need to select a factor that relates failure probability to fatality (or injury) probability. Such a factor will vary among components and will especially depend on the amount of time that personnel are around the components. The factor may also vary with failure type. You will need to modify the safety limits in the spreadsheet(s) to accommodate the factor(s). This approach is more complicated, but it can add value if the year of maximum NPV is not being achieved because of safety limit constraint. However, you should learn the process before you try to add this consideration.

10.1.3 □ Determining Projected Unit Capacity Factor (Demand) and Forced Outage Consequence

You are importing one element of risk—the failure probability information that you developed by using previous procedures. You made a preliminary risk-ranking by using production (MWH) loss estimates. Now you need some more detailed consequence information to complete the risk basis for the calculations.

The production plan for a piece of equipment is an important economic failure consequence component. The average breakdown consequence cost for lightly or infrequently loaded equipment is generally much less than the breakdown consequence cost for equipment that is used a large percentage of the time. The predicted loading for a piece of equipment is controlled by the equipment's function in the system and by the system production plan.

The information source for system demand is in the corporate strategic production plan for the system. This plan is usually produced by the corporate production planning group. You should identify the units that the components that you have selected for analysis will affect, then contact the production planning group to determine the year-by-year projected demand for each such unit. Get the "projected capacity factor," which is the percent of time at full production for each unit for each of the next 20 years. If, when obtaining this number, you discover that the plan does not exist for this long, then obtain the longest projection possible. This projection into the future is very important because the economic consequence of a decision is often controlled by the economic consequences of future failures.

When you know the projected unit capacity factors, you can use engineering knowledge of system function to determine the projected "capacity factors" for each selected component. You can usually calculate these factors by multiplying the percentage of the time that the component needs to be functioning by the unit capacity factor. For example: a boiler with a capacity factor of 85% requires the superheater tubes to function 100% of the time without a major leak. In this case, the capacity factor for the particular component, the superheater tubes, is $1.00 \times 0.85 = 85\%$. More complex system-component interactions may require a reliability block diagram or even fault tree analysis. See Section 8. This percentage of demand, or component capacity factor, is really the probability of need for the component.

The next task is to calculate the year-by-year incremental consequence costs for down time or a forced outage caused by each component. You will need, for each unit, the year-by-year differential cost per unit of production output for replacement product (energy) from the next less efficient unit in the system (dispatch) or from outside purchase. In a competitive environment, this would be called the lost opportunity cost per unit of production. This information, the incremental replacement product (energy) value, will normally be obtained from the same source as the unit capacity factor information. You will then know the incremental cost, per unit of production capacity, of a forced outage for the unit. You can use this information to calculate the economic failure consequence per hour for each component as follows:

$$CEFC/Hour = UC \times IFCC \times CCF$$

Where:

$CEFC/Hour$ = Component Economic Failure Consequence per Hour

UC = Unit Capacity

$IFCC$ = Incremental Failure Cost Consequence

CCF = Component Capacity Factor

For example, for a superheater tube,

$$CEFC/Hour = UMDC \times IREV \times CCF$$

Where:

$CEFC/Hour$ = Component Economic Failure Consequence per Hour

$UMDC$ = Unit Maximum Dependable Capacity

$IREV$ = Incremental Replacement Energy Value

CCF = Component Capacity Factor (0 to 1)

This calculation provides the consequential cost for each hour that the selected component causes the unit to be down. In the example that follows (See Section 10.3.1), this is the Forced Outage Cost for each hour the unit is down because of a superheater tube failure.

10.1.4 □ Knowing How to Maximize Value to the Company

There is a lot of emphasis today on "value added." The reason for this is, of course, that most companies are stockholder owned. Their objective is to maximize the return on each invested dollar. Maintenance is an area of particular concern, because maintenance dollars do not directly produce output. Maintenance is critically important, yet, spending a maintenance dollar in a manner that does not somehow provide maximum return is not making the best use of stockholder funds nor company resources.

Using Net Present Value (NPV) as a decision making criterion is a way to achieve this "value added" objective. NPV not only considers the investment of the resource but also accounts for the cash flow consequences over the service life of the equipment. This is why the major textbooks on basic corporate finance suggest maximizing NPV as the decision making criterion for investments. See (Brealey, 1991). Maintenance expenditures are investments that are made to receive the benefit of production return in the near, intermediate and long term.

NPV puts all corporate entities on a level playing field with a common communication medium for the corporate decision maker. This is why it is so useful in communications and decision making. The playing field is leveled because the units of measure are the same for any aspect of the company operation, including maintenance. Maintenance engineering proposal benefits, together with engineering analysis results, have been difficult to quantify for the decision maker. Concentrating on conversion of maintenance and engineering effects into cash flow and NPV terms will allow these to compete with other parts of the company in a clear manner. The common communication medium of dollars ties directly to the language of the decision maker and to the stockholders.

Optimization creates savings by looking at the long range life of the component, not just the short range view of the decision to be made.

The real plan is to operate these components for a number of years, therefore, looking at the financial effects over this whole time frame is most realistic.

There is a tendency to use payback periods to make project decisions. The difficulty with that approach is that it only looks at the short range. A financial process that also considers the long range benefit of a project provides a more realistic view of financial considerations over the service life of the component. This longer-range view occurs when NPV is used as the optimization criterion. For that reason, NPV will be used here. If another financial criterion is required, then the financial portion of the referenced spreadsheets can be modified to accommodate it.

The short term effects are accounted for in this process by annually updating the input data and then rerunning the optimization. This procedure provides the benefits of a realistic long range view over the service life of the component by, in effect, implementing the plan one year at a time. This one-year-at-a-time reoptimization accounts for the short term changes and effects that may occur and which need to be considered in any decision-making process.

10.2 □ ANALYTICAL APPROACH

One of the most important characteristics of the analytical approach to maintenance optimization is its use of NPV as the decision-making criterion in the world of maintenance decision making. The key to smoothly and successfully integrating maintenance and engineering decision making into mainstream corporate decision making is the calculation of NPV by including lost production costs in the failure consequence cost calculations. This is because the corporate decision making criteria are being used to express the need for maintenance resources. You can usually obtain the costs that you need for your calculations from the company production planning function. See Section 10.1.3. You might need to work through your manager to find where in the corporation this function resides.

You should use the corporate value measure (NPV) in your calculations but, especially at first, you should keep all the elements of your maintenance optimization decision model within the engineering, maintenance and operations world. Financial decision models usually include revenue information in their NPV calculations. By focusing the maintenance decision and its value structure in the engineering, maintenance and operations parts of the company, you are less likely to cause interdepartmental political problems, which, during the early stages of application of this approach, can create destructive conflicts. You can use lost projection costs (and projected capacity factors) to ensure that your model "slots in" to the corporate decision making process. Using revenue information, in addition to possibly causing conflict, adds complications that you do not need during the initial stages of your project.

The specific elements of the optimization model are:

1. The replacement or repair cost of the component(s) of concern
2. Realistic constraints such as annual budget limits
3. The production loss consequences if the project is not implemented
4. The production loss consequences if the project is implemented
5. The facility safety constrain limit

Items 1 and 4 are the project initiation costs. Item 3, the consequence that is avoided by performing the project, is the gross project benefit. The budget limit keeps the model grounded in reality by disqualifying "optimal" strategies that are unaffordable. Item 5 keeps the analysis grounded in reality by locking out any strategy that compromises safety.

The reasoning behind the model is this:

- If the proposed project is not implemented, there will be certain production losses, e.g., a "Do Nothing Different" or "Base Case" cost. Let this be C_B.
- If the proposed project is implemented, we will expect lower total production losses, e.g., a "With Project" or "Alternative Case" cost. Let this be C_A.
- If the proposed project is implemented, we will also pay the project cost. Let this be C_P.
- If the sum of the cash flows for the project implementation (replacement or repair) cost and the Alternative Case production loss cost is greater than the Base Case production loss cost cash flow, then a negative NPV will result. In other words,

$$NPV = C_B - (C_P + C_A) < 0.$$

We would decide not to implement the project.

- If the reverse is true, that the Base Case production loss cost cash flow is greater than the project implementation (replacement or repair) cost plus the Alternative Case production loss cost, then the NPV will be positive. In other words,

$$NPV = C_B - (C_P + C_A) > 0.$$

We would support this project.

In any case, the decision is constrained by projected maintenance budget fund availability during the implementation year(s) of the project and by the safety limit.

To illustrate these concepts, let's look at a decision that I might face at home:

- I have a car that has a rear tire that has a moderate amount of tread left. It is not a front tire, so I am not greatly concerned that a blow out will cause the car to lose control, but I am concerned that I might be delayed by a flat tire while I am traveling.
- I am considering recapping the tire at an expense of $50.
- If I continue to run the tire, my estimated probability of a flat is 50% this year and 90% next year.
- If I recap the tire, my estimated probability of a flat is 10% this year and 20% next year.
- My consequence of having a flat is 4 hours off of work at $25 per hour. Also, I have a daily medical treatment for which I will be charged $105 if I cancel without providing 24 hours' notice. I am not able to change the tire myself and will depend on a road service that will charge me $50. This is a total consequence of $255 if I have a flat. These assumptions are based upon the knowledge that the vast majority of my driving takes place within the city.

Do I recap the tire and, if so, should I do it this year or next year? The NPV calculations, without accounting for taxes or the time value of money, are tabulated in Figures 10.2.1 and 10.2.2.

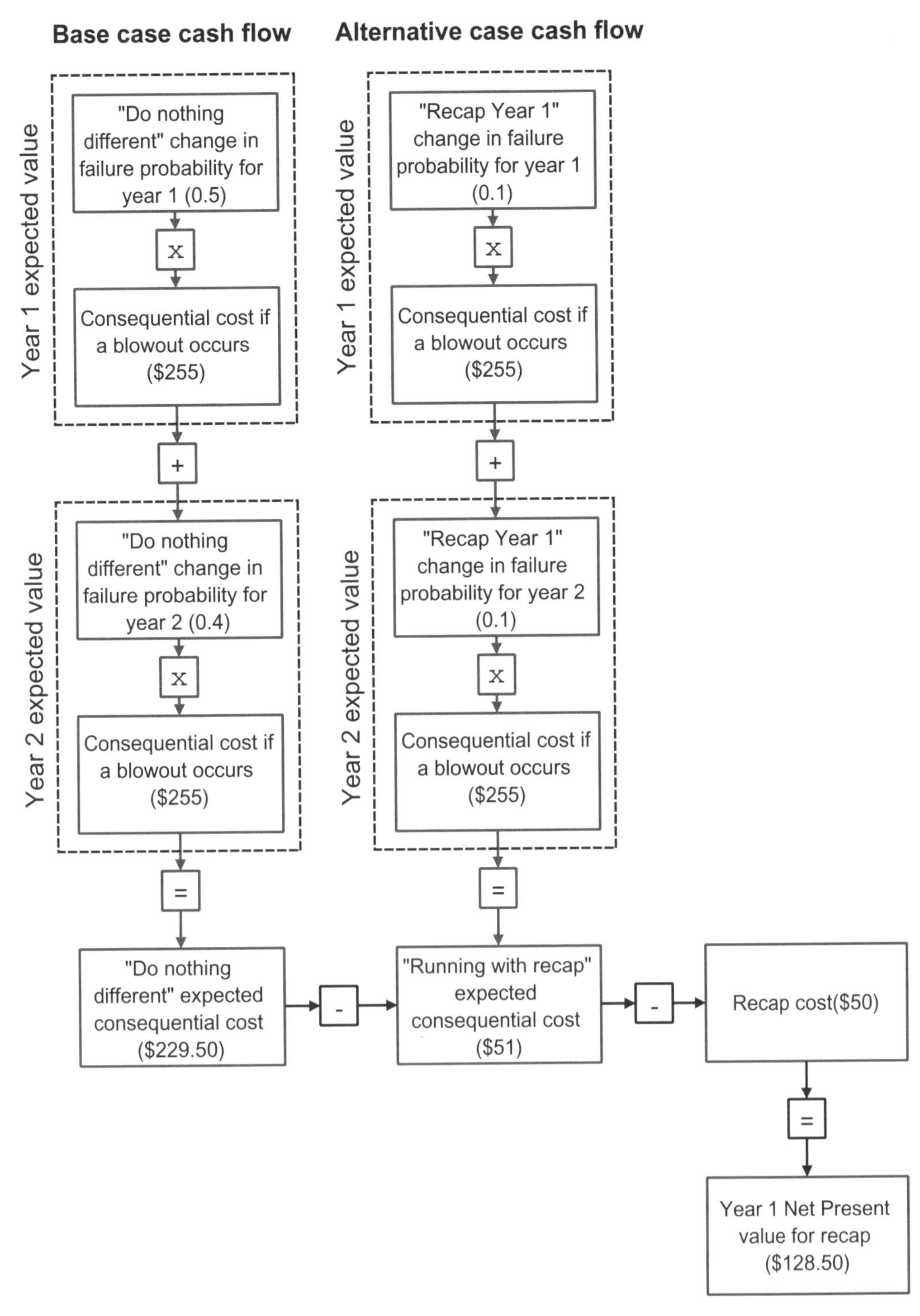

Figure 10.2.1 NPV Calculation for "Do Nothing" vs. "Recap in Year 1"

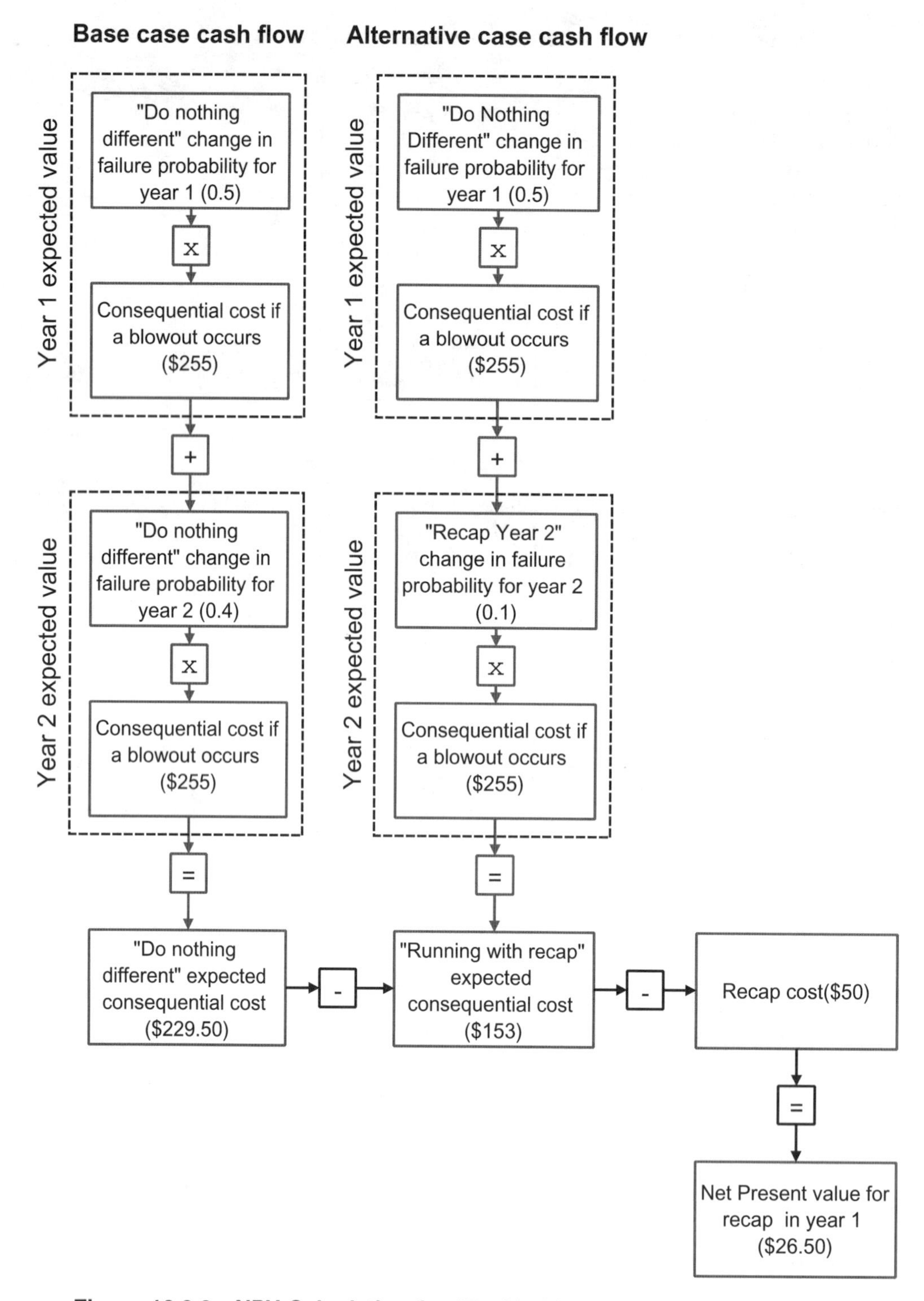

Figure 10.2.2 NPV Calculation for "Do Nothing" vs. "Recap in Year 2"

We want to choose the year that produces the highest positive NPV. We would gain by choosing either year because both produce positive NPVs, but the optimum year to recap the rear tire is the first year at an NPV of $128.50.

An optimizing model can perform this process for several projects simultaneously. The model helps us to chose projects that produce positive NPVs and to schedule those projects so that the overall NPV is maximized. This is the purpose of the optimization. The resulting prioritized list of project implementation years is the timing that will realize this highest NPV. The change in the overall NPV by optimization is the NPV produced by the whole process.

The model that you use must reflect the way that your company has decided to do business or wants to do business. The example spreadsheet on the Handbook CD-ROM, MANTOP™, has a conventional approach that uses after tax, discounted cash flows to produce NPV following (Brealey, 1991). It is important that you seek agreement with your company economic evaluation department that this example spreadsheet reflects the way that your company does business. If it does not, then ask this department to help you to modify this spreadsheet so that it does reflect your company's way of conducting these analyses. Your supervisor can help you get in touch with this department.

10.2.1 □ The Model Development Thought Process

The decision model is best illustrated by the decision analysis influence diagram. An influence diagram graphically represents the relationships that link a decision with its outcomes and their consequences.

For optimization of multiple components, the purpose of the influence diagram is to link the engineering world, expressed as failure probability, and the financial world, expressed as NPV. Engineers have often had difficulty expressing the need for financial resources in a format that communicated with the financial world and with the decision maker. The diagram itself is a robust communication tool that explains the make up of the decision and what influences it. Because the influence diagram contains no numbers or equations, it can be used to communicate with audiences from plant personnel to senior executives for buy-in.

The necessary nodes between the decision and the financial outcome are the consequence and relationship information that links them to form the complete model. When you construct the diagram, you start with the outcome node (NPV) and ask, "What do I need to know to calculate or determine NPV?" You continue to ask this question at each successive node, adding layers of nodes, moving from right to left, until you arrive at the decision that needs to be made. The engineering state of the component, as expressed by the failure probability, provides a natural way to incorporate the expected performance of the component as a result of the decision alternatives. The relationships that convert this failure probability into the basic components of NPV are what make up the bridge from engineering to finance.

Figure 10.2.1.1 is an influence diagram for the previous tire example. Note how NPV is on the right as the financial outcome. Leading to this node are financial consequences, "Cost of Road Service" and "Cost of Lost Work and Broken Medical Appointment." However, these values do not directly input to NPV because these magnitudes assume that the consequence definitely occurs. The real question is, "What are the expected magnitudes of these losses?"

We can usually make decisions when we have an idea of what to expect. To get the expected value of the loss, we multiply the total magnitude of the loss times the chance of the "Consequence." This "expected consequence" gives us the probability-weighted loss given the chance of the consequence, or the expected value. This term is commonly used in financial analysis. In engineering, we call this risk. By any name, this parameter is the average expectation of loss given the decision that we choose. The chance of the "Consequence" is "probability of tire blowout," which is influenced by the decision to either "Run" or "Recap" the tire. These nodes are the engineering part of the model. Note how the chance of "Consequence" times the total magnitude of the loss links the engineering part of the model to the financial part of the model in the "Expected Consequential Cost" nodes. This is where the link between engineering and finance occurs.

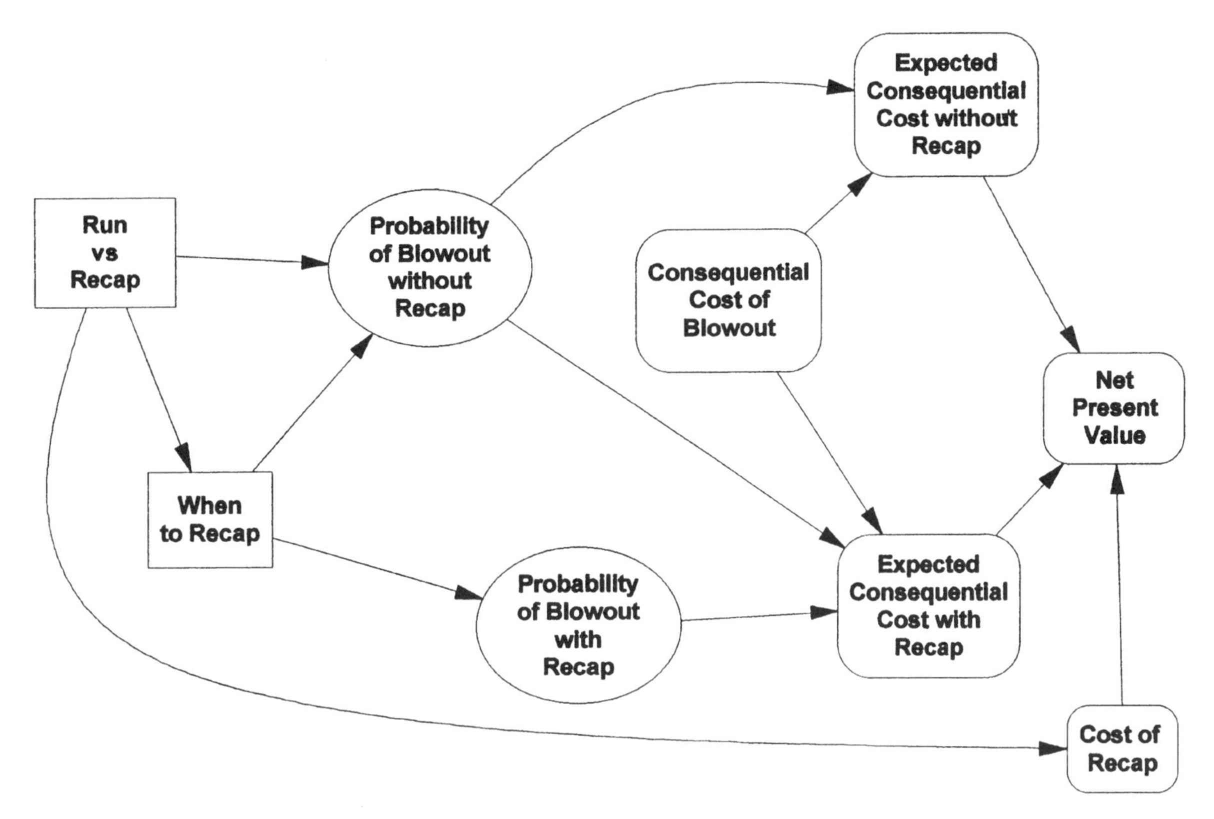

Figure 10.2.1.1 Decision Analysis Diagram for the Tire Recap Example

10.2.2 □ Developing the Decision Model

To illustrate decision model development, we will construct the diagram that is shown in Figure 10.2.2.1. We will start with the ultimate goal; to maximize NPV. On the far right, we draw a node to represent the NPV calculation. We then ask ourselves, "What do I need to know in the predictive maintenance arena to calculate NPV?" The answers are:

1. Total Discounted Lost Production without Project Implementation,
2. Total Discounted Project Cost - (Expense and Capitalized) and
3. Total Discounted Lost Production with Project Implementation.

Each of these costs become calculation nodes in the influence diagram. They are drawn immediately to the left of the NPV node and linked to it. The word "project" can be synonymous with "component" in the sense that we are considering some maintenance action, which we might call a project, on the component.

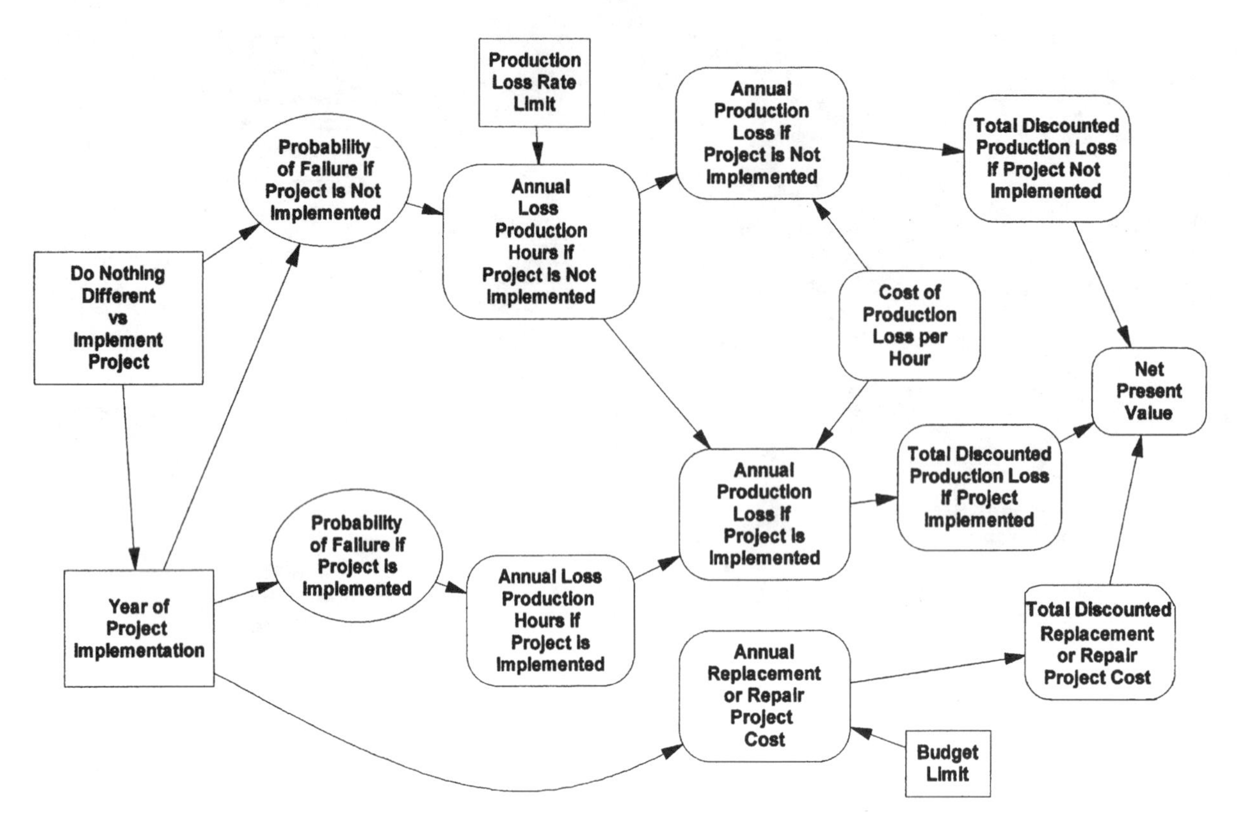

Figure 10.2.2.1 The Maintenance Optimization Decision Analysis Influence Diagram for Reliability

Then, for each of these costs nodes, we ask, "What do I need to know to calculate this node?" For each of these nodes we need to know the following nodes on a annualized basis:

1. Annual Discounted Lost Production Cost without Project Implementation
2. Annual Discounted Project Cost - (Expensed and Capitalized) and
3. Annual Discounted Lost Production Cost with Project Implementation

Now consider the next layer of nodes to the left. For the Lost Production consequence cost we need to know:

1. Cost per Hour of Lost Production
2. The number of loss production hours per year given the project is implemented and

3. The number of loss production hours per year given the project is not implemented

For the cost of the project we need to know, besides the cost of the project itself:

1. The year the project is to be implemented, to account for the time value of money and
2. The annual budget limit constraint for maintenance expenditures.

Each of these become calculation or information nodes in the influence diagram. They are placed immediately to the left of the associated node and linked to it.

Then, for each of these nodes we ask, "What do I need to know to calculate this node?" For the number of hours per year given no project implementation, we need to know:

1. The number of hours down given a failure or, if a multiple element component, the number of elements and the number of hours down if an element fails and
2. The change in failure probability each year given the project is not implemented.

Then for each of these nodes we ask, "What do I need to know to calculate this node?" For the number of Lost Production hours per year given project implementation, I need to know:

1. The number of hours down given a failure or, if a multiple element component, the number of elements and the number of hours down if an element fails,
2. The change in failure probability each year given the project is implemented,
3. The change in failure probability each year given the project is not implemented to account for the consequence of project delay and
4. The year that the project is implemented.

What we need to know now is the failure probability of the particular component versus time with and without project implementation. This

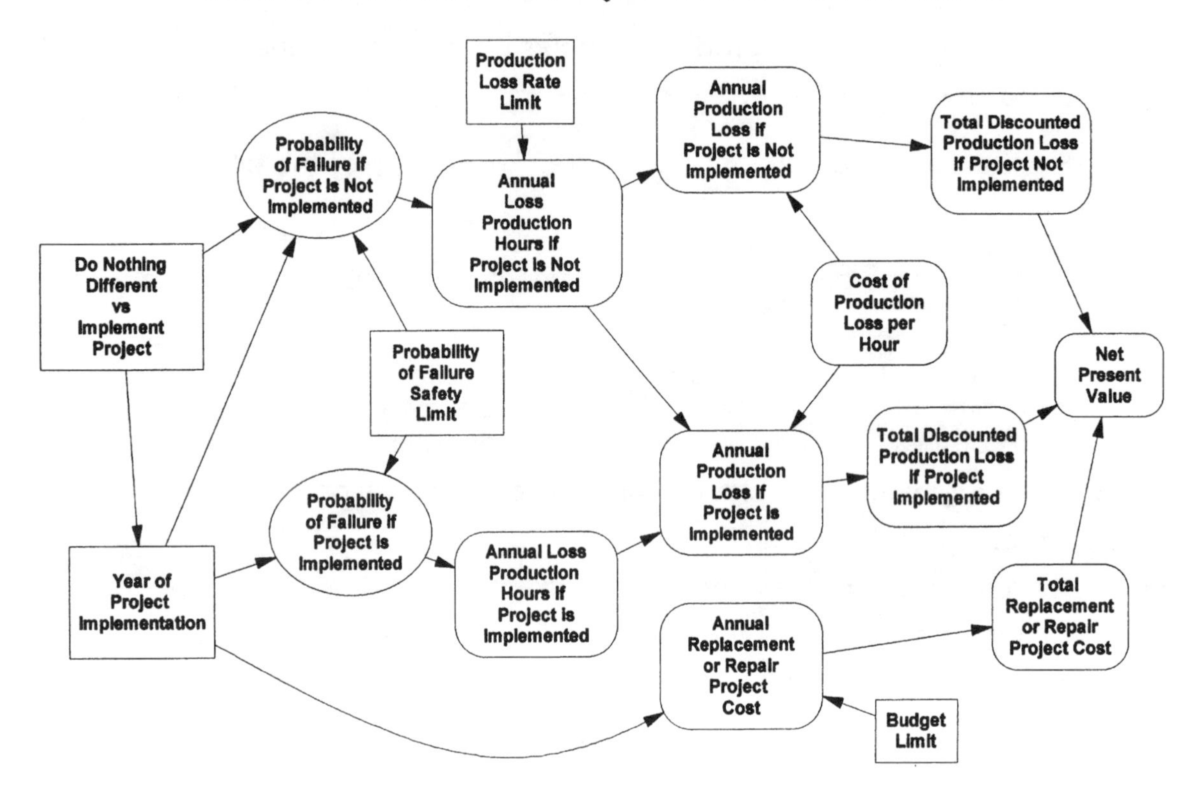

Figure 10.2.2.2 The Maintenance Optimization Decision Analysis Influence Diagram with Probability of Failure Safety Limit Constraint

comes from the methods that were described in Section 9. The only other thing that the model requires is the year of project implementation. This already exists in the project cost leg of the influence diagram.

In order to provide a Production Loss Rate Limit constraint on the problem to represent a corporate policy performance index, we need to know the Production Loss Rate (for a fossil utility this is Forced Outage Rate Limit). To know that, we need to know the expected service hours at full capacity as well as the number of hours down that a failure or multiple failures will cause, given the project implementation decision year.

A probability of failure safety limit constraint on the problem represents another corporate policy. This constraint is applied to the "alternative" case or "with project implementation" probability by year given the project implementation decision year. Any year in which that limit is exceeded is one year beyond the latest year that the project can be delayed. The influence diagram with this limit is shown in Figure 10.2.2.2.

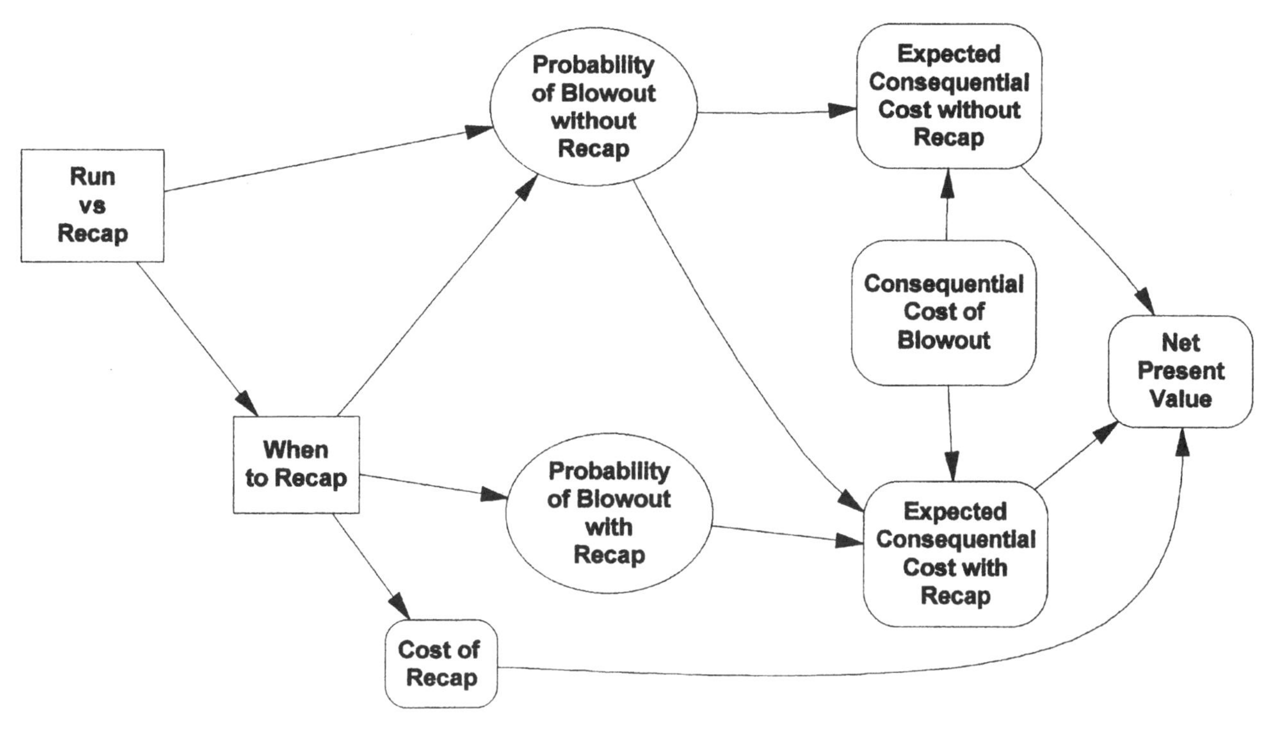

Figure 10.2.3.1 Time-Ordered Influence Diagram for the Tire Example

We know that we have completed the diagram because we have to go outside the diagram to get the uncertainty for the failure probability and the decision of project implementation year, budget and forced outage limit constraints. An influence diagram is complete when all its nodes that are not calculated within the diagram require information from outside the diagram or are decisions.

This diagram puts the details on the general concept model that was discussed in its five elements in Section 10.2.

10.2.3 □ Constructing the Decision Tree from the Influence Diagram

Converting the influence diagram into a decision tree is best described as moving the influence diagram nodes around until the nodes are nearly in time order while maintaining the interconnecting node arrows so that the relationship information is not lost.

For the tire example, notice in Figure 10.2.3.1 how the nodes have been moved around to reflect time order while maintaining the interrelationship arrows. At this point in the construction of a decision model, we would usually construct a decision tree. However, an optimization decision model is a special class of decision model in which we use an iterative process to optimize the decision criteria. The time node in a time-varying problem significantly multiplies the number of tree branches. Because of this complexity, the optimization-type decision model is not well represented by a decision tree. Such a model is better represented by a time-ordered influence diagram as is shown in Figure 10.2.3.1.

Now, taking the previous maintenance optimization influence diagram for reliability from Figure 10.2.2.2, we have taken the nodes and moved them around in Figure 10.2.3.2 to account for time ordering.

10.2.4 □ Using the Influence Diagram and Decision Tree to Construct the Optimization Model

The spreadsheet model is constructed by formatting the spreadsheet in the same time order as the influence diagram and the decision tree. The spreadsheet model must contain the elements of:

1. Component identification, to track the decision on each component
2. The year of action to be taken for each component
3. The failure probability with time for each component as a consequence of the decision action year
4. The relationships that tie the action year and failure probability to financial terms
5. The outcome expressed as NPV.

To construct the spreadsheet, we take the above spreadsheet elements and start laying them out from top to bottom in time order down the spreadsheet. We do this because, if we spread the sheet from left to right, we would require greater movement to see the same portion of the spreadsheet, because the cells are wider than they are tall. The supporting tables for calculation purposes will be placed to the right of their respective elements across the spreadsheet. This pattern does not have to be followed strictly, but it is a good way to help you start to develop your own system.

Load the file MANTOP.XLS into EXCEL from the Handbook CD-ROM

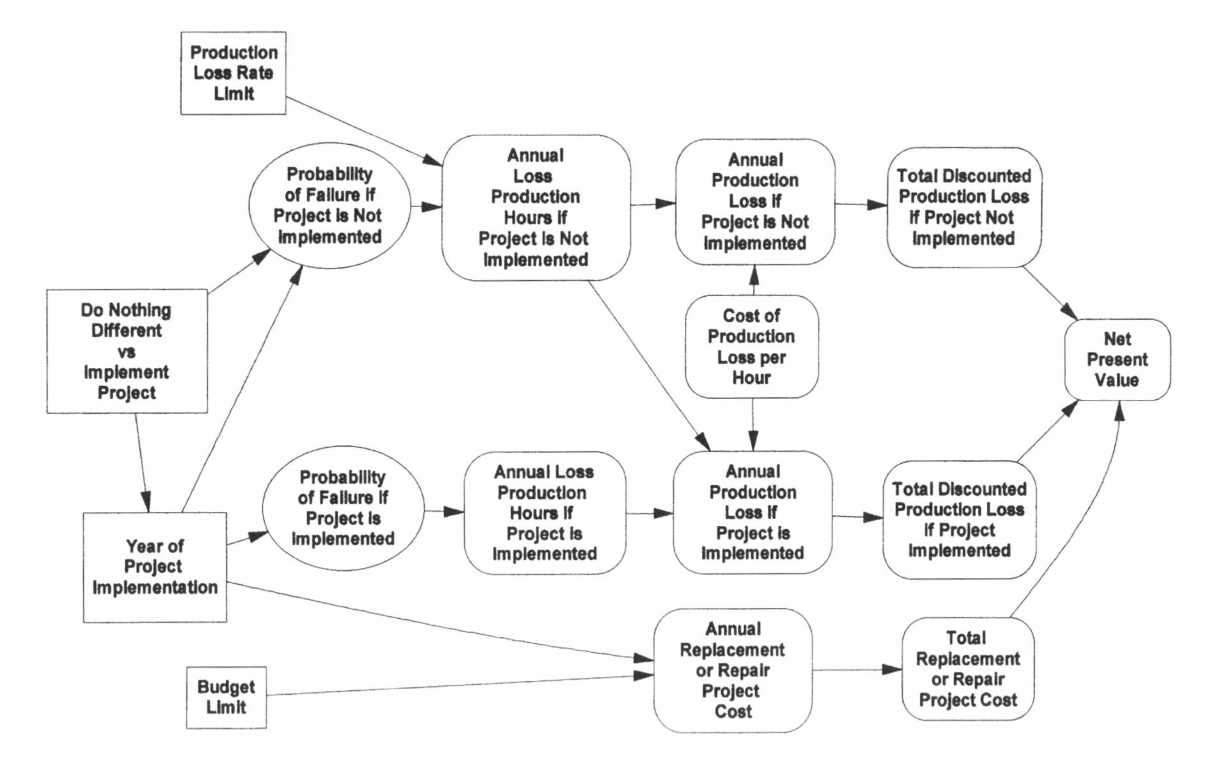

Figure 10.2.3.2 Time Ordered Influence Diagram for the Maintenance Optimization Model

and copy it to your working directory. See Figure 10.2.4.1. Notice how the table headings indicate the order of the basic elements down the spreadsheet that are present in the time-ordered influence diagram.

10.2.5 □ The Components of the Spreadsheet Model

In the manipulated influence diagram in Figure 10.2.3.2, the first node on the left is the Year of Project Implementation. Associated with this is the project cost. In the spreadsheet, the first boldfaced type heading indicates where this element of the spreadsheet is located. Next is the Component Failure Probability if the Project is not Implemented and to the right the Component Failure Probability if the Project is Implemented. Note that the probability as a function of time is laid out to the right. This is necessary for the model because this is a time-based decision.

The Loss of Production Hours if the Project is Implemented and Loss of Production Hours if the Project is Not Implemented are the next elements in the influence diagram. They are shown as the next elements down in

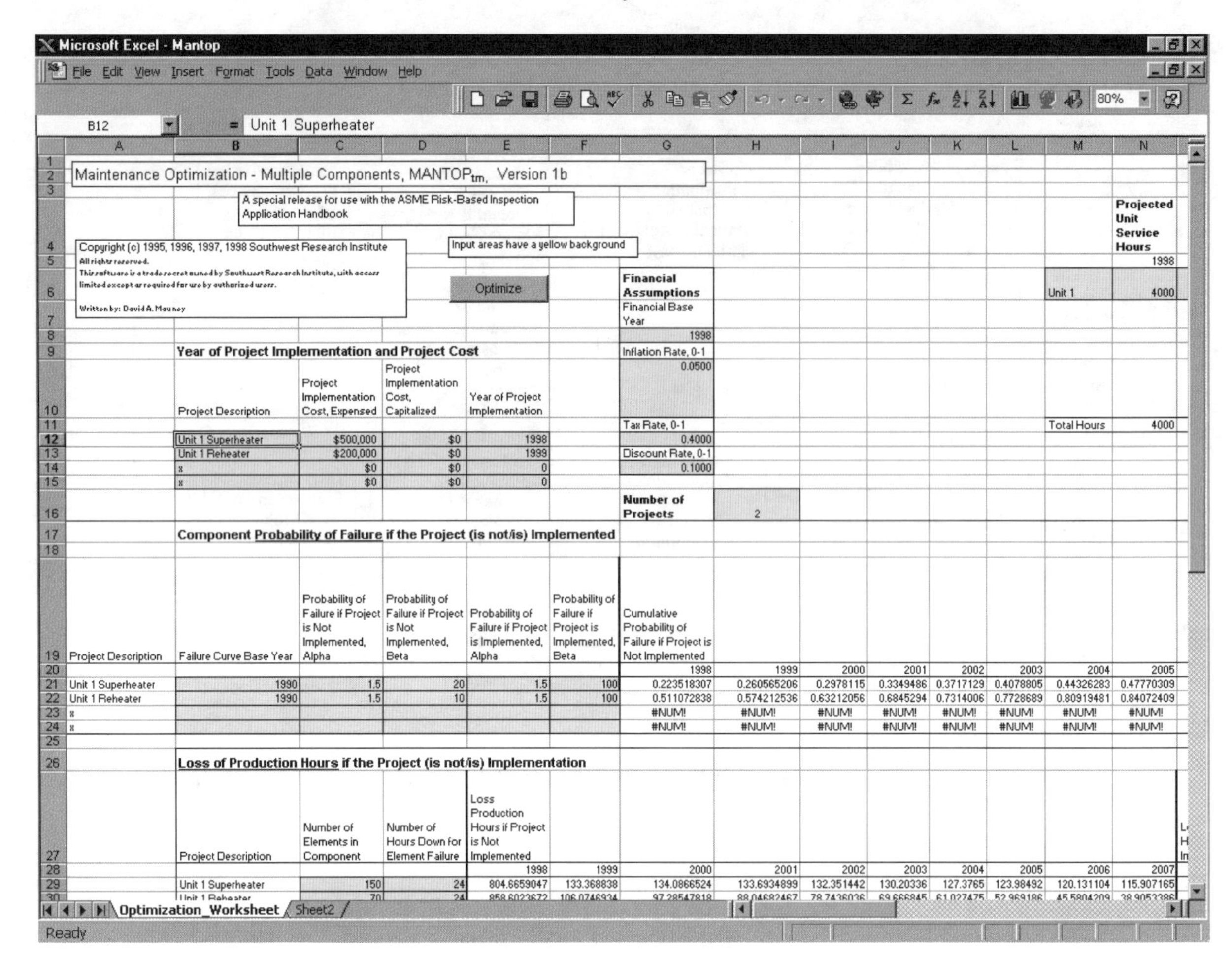

Figure 10.2.4.1 Multiple Component Ranking Based on Optimization Spreadsheet MANTOP.XLS

the spreadsheet. Again, the value change of each node with time is laid out to the right. The annual sum across all projects for each year is laid out further to the right. Similarly, the Production Loss in Dollars if the Project is Implemented and the Production Loss in Dollars if it is not Implemented for each year is the next spreadsheet element down. The annual sum across all projects for each year is further to the right.

The last table in the spreadsheet is a summary table that ties together Annual Production Loss with Project and Annual Production Loss without Project, Annual Project Cost, Budget Limit, Forced Outage Rate Limit, Total Cost and NPV. This arrangement of the summary table allows the annual cash flows for each path that is shown in the influence diagram to be displayed. Note that the spreadsheet flows from top to bottom and ends in the ultimate outcome, NPV.

Now, to use the influence diagram to show the relationships of the parts of the spreadsheet, it is most helpful to start at NPV, just as we did when we originally constructed the influence diagram. The NPV is the sum of all the Total Discounted Cash Flows just above it as was indicated in the influence diagram. The totals are the sum of each of the Annual Discounted Cash Flows above it.

For each year, the Annual Discounted Value is a function of the Annual Cost Value immediately to its left. The discounting equations appear in the Annual Discounted Cost Columns, except for the Annual Discounted Project Cost, Capitalized. This is processed through the Tax Effect of Depreciation Table, which is located to the right of the summary table.

The Annual Budget Limit and Production Loss Forced Outage Rate Limit Constraints are just to the right of their respective nodes. The Annual Cost for each cash flow stream is derived from the annual sum line for each year in the cost table above the summary table. The exception to this is the project cost, which is a summary by year of the first element table for Project Cost dependent on Implementation Year of the project.

The Annual Production Loss Cost of the projects is derived from the Loss Production Hours element table for each project for each year. For both the Projection Loss Cost and Hours element tables there is a link to the Year of Project Implementation for each project by cell references.

Finally, the Production Loss Hours element table is linked to the Component Failure probability element table whether or not the project is implemented. Although the tables show the probability of failure as cumulative, the annual change in probability of failure is calculated in the Annual Loss Production Hours tables.

10.2.6 □ How the Optimization Model Works

It is easiest to describe how the spreadsheet performs an optimization by referring back to the influence diagram in Figure 10.2.3.2. Note that here we have all the elements of the model connected together. Remember that we also have a spreadsheet that performs all the calculations that correspond to the relationships that are shown in the influence diagram.

The optimal NPV is determined by finding the Year of Project Implementation that will cause a higher NPV than any other year. Each time a selected year produces a higher NPV than the previously-selected year, a check has to be made to ensure that neither of the constraints are exceeded for any year. The year that causes the highest NPV without exceeding any constraints is the optimum year for project implementation.

In the spreadsheet, the optimization is expanded into multiple projects. Here, the process is the same, except that the combination of years is sought that will produce the maximum NPV without exceeding any of the constraints in any year. Once this combination is found, the projects as a group will have been optimally prioritized, each with its individual timing.

10.2.7 □ How to Optimize the Spreadsheet Model

It is best to provide spreadsheet inputs that are generic to your system so that the spreadsheet can be modified for various project scenarios or groupings.

1. For a reliability-related set of components, copy the example spreadsheet MANTOP.XLS from the Handbook CD-ROM to your working directory. Copy MANTOPS.XLS if you want to optimize a safety-related set of components.
2. Save the sheet in a directory under a name that is meaningful to you, however, you MUST use the "8.3" naming rules. In other words, the spreadsheet must be named "XXXXXXXX.xls," where each "X" is a number, letter, underscore or one of these characters: ^ $ ~ ! # % & - { } ( ) @ '
3. Click on the menu item [Tools] and click on [Solver] if it is there. When the Solver box opens, push the [Close] button. You have just loaded Solver. If [Solver] is not listed in the menu, click on [Add-Ins], look for the checkbox [Solver Add-In] and check it off. Then push the [OK] button and continue. If the [Solver Add-In] check box is not in the [Tools] menu, you will need to run Excel Setup to install Solver and then follow the proceeding procedure.
4. Go to cell G8 and type in the financial base year (usually the

present year) for the analysis, and also the inflation rate, composite tax rate and discount rate where indicated.

5. In Cell M6, type in the unit name and number for the unit for which you will enter components. It is usually best to put a unit name, a space and the unit number, e.g., Franklin 3.
6. In Cell N6 : W6, type in the projected number of service hours by year for the unit.
7. Save this spreadsheet under a new name that is unique to your system base spreadsheet for this unit.
8. Go to cell B12 : B15 and type in the project description and cost information. Make sure that each project descriptor is unique.
9. Type in the project capitalized and expensed cost information. To initialize the optimizer, leave zeros in the shaded area for years of implementation. If you are not going to fill all the available project lines with projects, then place an "x" in the unused project description fields (B12 : B15). If you have loaded MANTOPS.XLS, enter the Probability of Failure Safety Limit for each project/component.
10. In cell H16, type in the number of projects for which you have entered data.
11. In B21 : F24, place the respective "implemented" and "not implemented" Weibull equation Alpha, Beta and the common base year values for each project. The cumulative probabilities to the right will automatically be calculated.
12. Beginning in cell C29 : D32, type in the number of elements for each component and the number of hours down for each failure of each component. Again, all tables to the right will automatically be calculated.
13. In cell C37 : D40, type the replacement energy values in the indicated units and in the next column, the maximum dependable capacity. Of course, each of these values is for the unit that is being considered.
14. In columns L44 : L53 and N44 : N53, type the annual budget limit and production loss rate limit constraint values, respectively. At this point, you have fully loaded the spreadsheet for optimization.

15. Save this spreadsheet under a new name so that you will have a pre-optimized version that you can call up for other analysis scenarios.
16. To optimize the spreadsheet click on the Optimize button. The optimization will begin.
17. After the macro runs for a while, it might stop and open a box entitled "Show Trial Solution" with a sub-message that states, "The maximum iterations limit was reached; continue anyway?" If you need to stop the program for some reason, press [Stop]; otherwise press [Continue]. After a while longer, a similar box might appear with the sub-message, "The maximum time limit was reached, continue anyway?" Again, if you need to stop the program then press [Stop]; otherwise press [Continue].
18. When the optimization run is complete, the "Solver Results" will pop-up. Leave the selection on "Keep Solver Solution" and press the [OK] button.
19. You can now examine the optimized spreadsheet. It is usually advisable to run Solver several times, each time leaving the years from the previous optimization. You can stop repeating the optimization when the years do not change. This is an iterative process; it sometimes takes Solver several runs to settle on the solution.

The result of the optimization will appear. Note that years have now been placed in the Year of Project Implementation column, E12:E15. These are the years that maximize the NPV, seen in cell H60, and stay within the annual budget limit and forced outage rate constraints.

During the optimization, the projects with the highest NPV are selected in sequence and the year for each project that results in the highest NPV in time while staying within the annual budget and forced outage rate limit constraint is determined. When it is complete, the Year of Project Implementation column provides a prioritization of the projects in time. Additional projects would be calculated the same way.

The optimization gives you a prioritized listing of project implementation years for the project mix and input conditions that will maximize the NPV and remain within the constraints. This is the project timing that will produce the maximum NPV. If you want to investigate the effect of deviation from the maximum combination, type a new year in the shaded area, E12 : E15, for the project of interest and the spreadsheet will recalculate and provide a new NPV in the summary table. The drop in NPV is the value loss for the deviation from the maximum-value-producing combination.

Remember that some choices might produce higher NPVs, but you will find that one or more constraints will be exceeded to accomplish that. If this happens, you can see where the constraint is exceeded by looking in the summary table, i.e. I44 : I53 plus K44 : K53 compared to L44 : L53 and M44 : M53 compared to N44 : N53.

With the same awareness of budget and forced outage rate limits possibly being exceeded, trial years can be typed in cells E12 : E15 without an optimization being conducted. This allows "what-if" questions to be addressed on timing of projects years and their effect on NPV.

10.3 □ EXAMPLES

The first example, a boiler tube optimization, is a simplified example that deals with economic optimization only. The rationale is that boiler tube failures that occur inside the boiler are not likely to be a safety issue. The second example, a piping optimization, does have safety implications and is, therefore, economically optimized but is restrained by a safety limit.

10.3.1 □ Optimization of Boiler Tubes

The previously demonstrated maintenance optimization model, MANTOP, for reliability-related components, has the following reinterpretations when it is applied to boiler tubes.

The relationships in the influence diagram are the same but the wording in some of the nodes is different. The diagram in Figure 10.3.1.1 is the same as the general diagram in Figure 10.2.3.2 except for some fossil utility-specific word changes.

Table 10.3.1.1 contains equivalent titles that relate the fossil plant-specific multi-component optimization model to the general multi-compo-

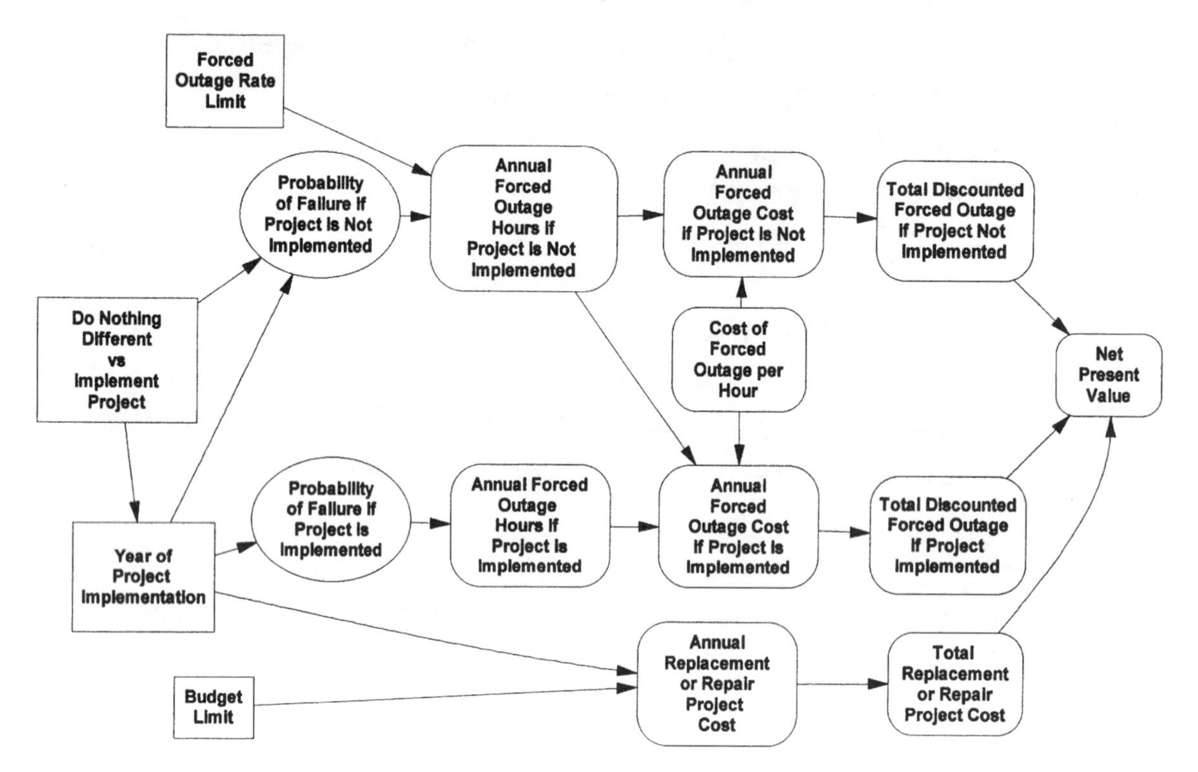

Figure 10.3.1.1 Time Ordered Influence Diagram for the Maintenance Optimization Model with Fossil specific Titling

nent optimization model. Note that when you load MANTOP™, the two components that are already shown are fossil plant tube components. This table is to help you to translate the general model titles to the fossil plant-specific application.

TABLE 10.3.1.1 EQUIVALENT LABELS FOR GENERAL MANTOP™ AND FOSSIL PLANT SPECIFIC MANTOP™

General MANTOP™	Fossil Plant Specific MANTOP™
Project	Fossil Plant Component
Loss of Production Hours	Forced Outage Hours
Production Loss in Dollars	Forced Outage Cost
Cost of Lost Production per Unit of Capacity per Hour	Replacement Energy Value
Maximum Consistent Unit Production Capacity	Unit Maximum Dependable Capacity
Annual Unavailability Rate	Forced Outage Rate
Annual Unavailability Limit	Forced Outage Rate Limit

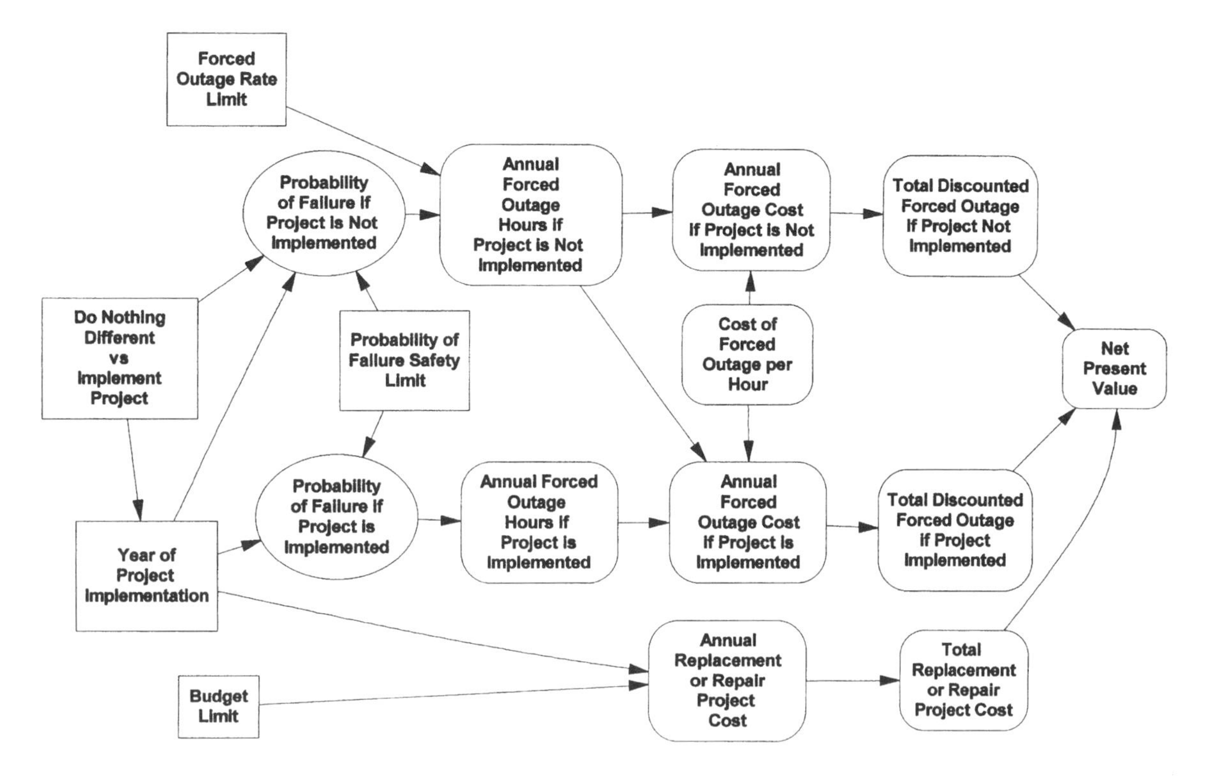

Figure 10.3.2.1 Time Ordered Influence Diagram for the Maintenance Optimization Model with Fossil specific Titling and Safety Constraint

10.3.2 □ Optimization of Piping Inspection

The previously demonstrated maintenance optimization model, MANTOPS, for safety-related components, has the following reinterpretations when it is applied to steam piping.

For safety related components like steam piping, you will need to slightly modify the influence diagram for the general case. Figure 10.3.2.1 shows how you might make the changes. This influence diagram shows the same relationships as does the general one in Figure 10.2.2.2, but the wording in some of the nodes is different. The diagram in Figure 10.3.2.1 also includes some fossil utility-specific word changes.

In the MANTOPS.XLS spreadsheet, the probability of failure safety limit in cells F12 through F15 is compared with the highest change in annual probability of failure for each component in cells AC67 through AC70 for the case where the project is implemented.

MULTIPLE COMPONENT RANKING BASED UPON OPTIMIZATION

NOTES:

SECTION 11

To perform this procedure, you will need:

Optimized inspection dates for the selected components. See Section 10.

To tailor this procedure to your facility, you will need:

- *The value measure that your facility uses for financial decision making. The example spreadsheet assumes maximum net present value (NPV). You can modify the NPV assumptions or you can modify the spreadsheet to calculate a different value measure.*
- *Projected loss opportunity cost for the unit(s) that contain the selected components.*

This procedure will develop inspection strategies for the selected component(s).

The next step will, in general, be sensitivity analysis. Or, you could select another "problem" and start the process over.

INSPECTION PROGRAM DEVELOPMENT 11

Now that you have prioritized the components in time and selected those in most immediate need of attention, you can focus on an inspection strategy for these components. You will have set the criteria that you will use to make inspection program development decisions in Section 10. You will have calculated the best time(s) to take action. Now, you will need to build the decision model that will link the component engineering information, expressed as failure probability versus time, to the other data for each inspection program decision scenario that you want to consider.

As we said before, this can be a very expensive and troublesome step. You will need information like nondestructive testing detection probabilities that practitioners might have difficulties providing. Worse, you might learn that you cannot reliably detect your key failure made at the level that you require. You will explore the distinction between inspections that make you feel good and inspections that actually are good.

For safety-related components, you should have safety-constrained the optimized schedule that you developed in the previous section. If you want to safety-constrain the inspection program, then you can use the optimizer in Section 10 as a pattern to modify the decision models in this Section and their use. Specifically, you can look for decisions that produce an inspection plan that results in a failure probability that is less than your failure probability safety limit but which gives the highest NPV within that limit.

11.1 □ STRATEGY TABLE DEVELOPMENT

A strategy table is a tool that will help you to logically construct an influence diagram. An influence diagram is a tool that, together with a decision tree, will help you to model the decision process. The influence diagram shows the relationships in a decision. A decision tree helps you to account for the time ordering of the decisions to be made and of the uncertainties in the decisions. As you gain experience in decision analysis, you may develop the skills and thought processes needed to construct an influence diagram, decision tree and then a decision model without having to construct a strategy table.

Your goal is to mathematically model the decision process. You will use the model to test different decision alternatives. Your decision model will calculate the expected values for various combinations of parameters, such as failure probability versus time. The calculations will account for the uncertainties in the inputs and specify the uncertainty in the calculated value. Finding the decision that is expected to maximize the value measure is the goal of the decision analysis for inspection program development.

Constructing a strategy table is the first step toward developing the decision model. You might think of the strategy table as a way to present your thoughts about what information needs to be included in the decision making process. It is a structured way to "brainstorm" your ideas so that you can lay out all the information that might be involved in the decision in an orderly format. One of the strengths of decision analysis is that it provides order and logic to the decision-making process.

Before we develop an inspection strategy, we will use a simple example to introduce you to the decision model development process. Our example deals with a badly worn tire on my car. I can choose to run the tire "as-is" instead of replacing it. This avoids buying a new tire today but exposes me to a higher probability of suffering one of these consequences:

1. Blowout in town = \$255 = \$105 for failing to give 24 hr notice for missing a daily medical appointment + \$100 for 4 hr lost work + \$50 for road repair service (my medical condition prevents me from changing the tire myself).
2. Blowout on the highway (out of town assumed) = \$340 = same as "in town"+ \$50 for an additional 2 hr lost work + \$35 for more costly road service.

(Both scenarios assume that the blowout will not result in an accident.)

Or, I can replace the tire with a new one and suffer spending $50 for the replacement now but reduce the probability of suffering a blowout and its consequences.

As in Section 10, the new tire could be purchased, the alternative case, to avoid the consequence of running the worn tire, the base case. This means that we are declaring the consequence of running the worn tire as a benefit and the cost of purchasing the tire and it's associated consequence as the cost. The outcomes I am estimating can only occur after either I actually buy a new tire or do not buy it and just keep running the old tire. We will walk through the decision process in detail.

11.1.1 □ Laying Out Strategy Categories

Several steps in the development of the tire replacement decision model are contained in the example spreadsheet TIRE.XLS. If you wish to follow the model construction steps, load Microsoft Excel and open the spreadsheet TIRE.XLS which is located on the Handbook CD-ROM.

Click on the tab, [Original Worksheet]. See Figure 11.1.1. The left side of the table has space for row headings for the different candidate inspection strategies that will be compared and for the damage mechanisms that will be addressed. The column headings across the table are for the categories of information that will be involved in the decision.

Choosing the ultimate strategy is what decision analysis is all about. The first step in laying out the contents of the decision model is to identify the information categories that might influence the decision, from the damage mechanisms through the failure mode and probability information all the way to the decision criteria. Think of a category as:

- A quantity that you need to know before you can estimate the economic decision criteria or
- A resource that you will need to choose before you can estimate the economic decision criteria or
- A value that you will need to calculate.

If you are developing an inspection strategy, the categories will in-

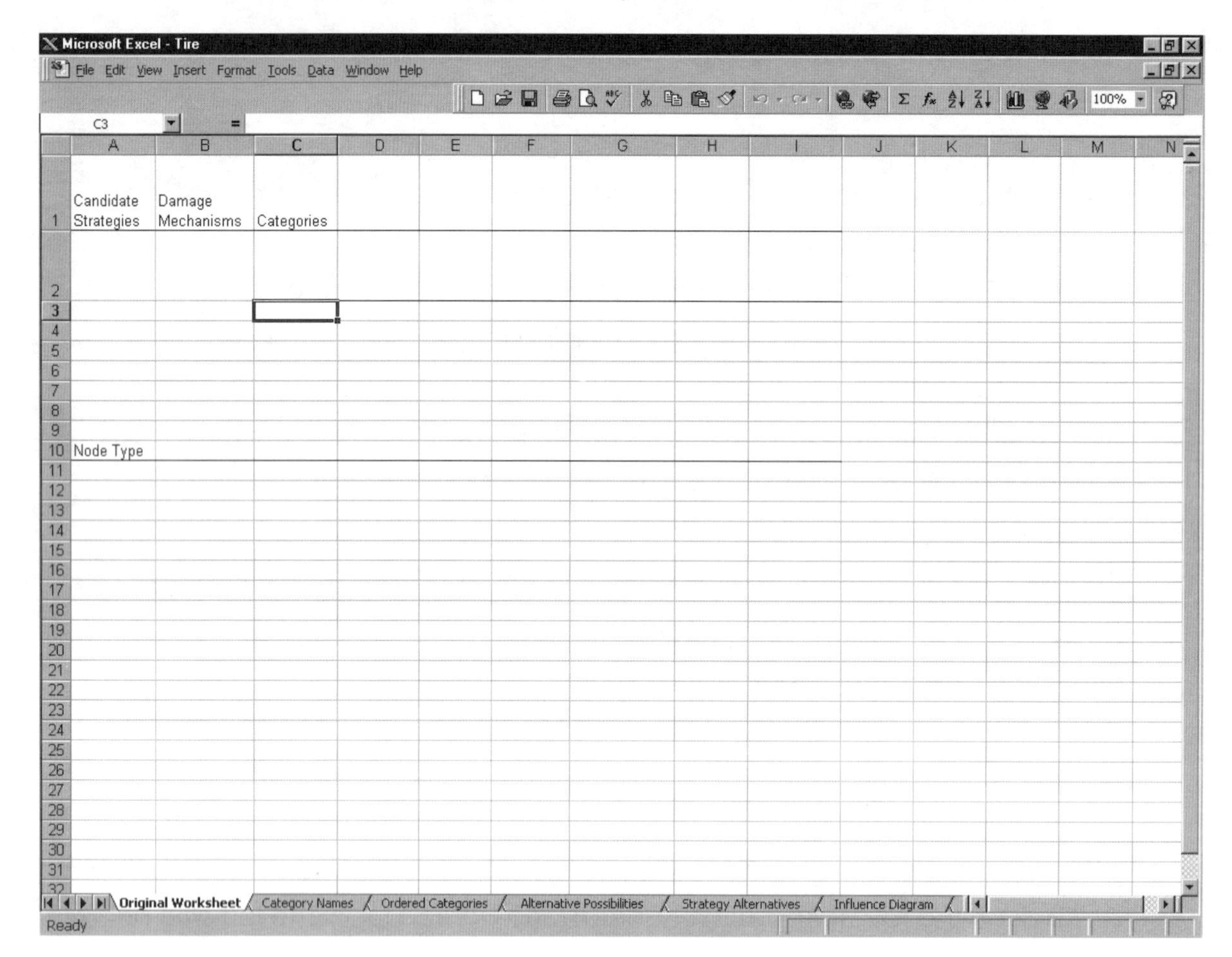

Figure 11.1.1 [Original Worksheet] in TIRE.XLS

clude decisions that you need to make and things that you need to do to perform the inspection. You can say that the categories are the items that will unfold after the decision is made and which will all link together to determine the magnitude of the economic criteria.

You can just list on a scratch piece of paper or "brainstorm" a list of the categories of information which you will need or the magnitude of which you will need to consider when you are making the decision. A list for a worn tire replacement decision might be:

- Net Present Value
- Risk of Blowout
- Tire Condition
- Probability of Blowout

- Consequential Cost of Blowout
- Car Use
- Tire Replacement Cost

[Category Names] in TIRE.XLS shows these categories entered in the worksheet from columns three over, starting with Net Present Value and continuing in "brainstorm (e.g., random) order." TIRE.XLS tab [Ordered Categories] shows these categories reordered from left to right in a logical time sequence.

11.1.2 □ Define the Candidate Strategies and Damage Mechanisms to be Addressed

The first column of the table lists the candidate strategy choices. In the case of the tire, the choices are "Run" or "Replace" (the old tire). The damage mechanism that drives the need for the strategy is (tire) "Wear," which is entered in the second column of the table. See TIRE.XLS tab [Ordered Categories].

11.1.3 □ Define Decision Node Types for Each Category

Look at each category and decide which type of decision model node that category represents. You have three choices:

1. Decision nodes involve choices between two or several alternatives that address the mechanism that you have selected.
2. Chance nodes represent categories with several possible outcomes.
3. Value nodes each contain a single value or a choice of values that is input from outside the model or calculated by the model.

In TIRE.XLS tab [Ordered Categories], the node types for each category have been defined and entered.

A hint on how to do this: Look at the type of information that you would put into each category. Consider, for example, the "Probability of Blowout" category. This node requires uncertain information, so it is a chance node. On the other hand, the "Tire Replacement Cost" requires numeric input, so it is a value node. "Candidate Strategies" requires a choice between strategies, so it is a decision node.

11.1.4 □ Provide Alternative Choices for Each Strategy Category

You have now defined the constituents of the decision. The next task will be to "flesh out" these constituents with alternatives to be considered with quantitative information.

Think about how each of these outcomes will have a certain chance of occurring. The probabilities of these occurring will be assessed later.

Look at each category, one at a time and start filling in the information that the node requires, as follows:

- A decision node, such as "Category Strategies," requires specific alternatives.
- A chance node, such as "Tire Condition," requires a list of outcomes.
- A value node requires a number. If the number will be calculated by the model, you may merely enter "calculate" for now.

The decision model is beginning to take shape. See TIRE.XLS tab [Alternatives Possibilities].

11.1.5 □ Link the Alternative Choices in Each Category to Form Strategies

Now we start linking things together to define the different strategies.

1. Look at TIRE.XLS tab [Strategy Alternatives]. The candidate strategies on the left make a natural starting point. The first candidate strategy is circled in green to help you to focus on the first candidate strategy.
2. For each candidate strategy, circle the applicable damage mechanism. In order, from left to right, circle the alternative choice or outcomes in each successive category that is associated with the candidate strategy/failure mechanism pair that you are considering. Remember that some alternatives apply only to the alternative directly to the left and some alternatives apply to all the alternatives. For example, the "Probability of Blowout" in each row only applies to the "Tire Condition" to its left. The "Car Use" values apply as a group to each candidate strategy. Be careful to draw the circles correctly in accordance with the intended relationships.
3. Go back and link the circled alternatives and possibilities with lines

that join adjacent categories across the table. This develops the "Run" strategy in green.

4. Performing the same process again but using red develops the "Replace" strategy.

The finished product will look like TIRE.XLS tab [Strategy Alternatives].

11.2 □ INFLUENCE DIAGRAM DEVELOPMENT

An influence diagram is a layout of the relationships that take the strategy alternatives through their uncertain outcomes and related consequences and value relationships to the estimate of the financial outcome. We will now construct the influence diagram for the tire replacement decision.

Probably the easiest way to do this is to take each element from the strategy table and place it in time order from left to right as unconnected nodes. We then start with the financial outcome, or value measure node, which we locate on the far right, and ask the question, "What do we need to know to calculate (in this case) NPV?" If the existing node to the left is all that we directly need, then we connect the two nodes with an arrow. If we need to know something else, or if we find there is an intermediate step between the existing node to the left and the NPV node, then we add or move node(s) as needed to assist in the relationship building or to provide information. When the required nodes are inserted and correctly ordered, we connect them with arrows. Note that the value nodes are green-bordered rectangles, the chance nodes are ellipses and the decision nodes are red-bordered rectangles.

We work through the diagram toward the left, taking each node and asking the question, "What do I need to know to calculate this node?" We then connect existing nodes or add new ones on the left and connect them with arrows until we have connected all the way back to a strategy alternative decision or have defined nodes for which the information will have to come from outside the process. The finished influence diagram is tab [Influence Diagram] in TIRE.XLS. Note that the "Strategy Decision" (labeled as such) is one terminal point for the diagram and that "Car Use," because it cannot be determined within the decision process, is the other.

11.3 □ DECISION TREE DEVELOPMENT

The schematic decision tree is a time-ordered description of all the choices in a decision model and of all the possible outcomes or consequences that each choice will produce. It is viewed in time order from left to right to represent the fact that we do not know results from choices until we implement them. Influence diagram and schematic decision tree construction force us to clarify the relationships in the decision model. The influence diagram establishes relationships between categories and the schematic decision tree clarifies the time order of the decision/consequence sequences among the categories and their alternatives.

All the decision and chance categories in the strategy table are to be represented in the decision tree. The best way to accomplish this is to take the time order of the categories from the strategy table. Our strategy table, Figure 10.2.2.5 and TIRE.XLS tab [Strategy Alternatives] has been time-ordered. Strategy tables are not necessarily time-ordered, however we find that this is a useful approach to use when we develop these diagrams. The following steps produced tab [Schematic Tree] in TIRE.XLS.

On the left is the strategy decision. Note that the two strategies are listed (in the blue background area) to start the tree. They are spaced to leave room for developing the other categories in the tree to the right.

The consequence for each decision, e.g. the condition of the tire we will be running under each scenario and the resulting failure probability, follows to the right under a red highlight. The heavy black line on the left of the red area shows by its height the strategy choices to which it applies.

Continuing to the right, the car use and its associated consequential cost of failure (blowout) applies equally to either decision. The duplicate "City" and "Highway" chance nodes and their consequences indicate this. The heavy black line ties the chance outcomes together.

This process is continued until all the decision and chance nodes are included.

11.4 □ SPREADSHEET CONSTRUCTION

The decision model spreadsheet is constructed by laying out the spreadsheet cells from left to right in the same time order as was used in the strategy table and decision tree. The equations within each cell will represent the relationships that are shown in the influence diagram. The multiple input values for each cell in the decision model are described in the decision tree. The spreadsheet model must contain the elements of:

1. The decision for each strategy
2. The uncertainties within each strategy
3. The computations that relate the decisions and uncertainty
4. The value node computations that relate the other nodes to the value criteria

To construct the spreadsheet, we lay out the categories that we have been using all along from left to right across the top. The supporting tables for calculation and data input will be placed below their respective categories. The decision model appears in TIRE.XLS tab [Decision Model]. Notice how the column headings across the top of the spreadsheet follow the order of the basic categories that appeared in the strategy tables. (The colored cells in the spreadsheet contain embedded notes that describe the cell contents. Briefly hold the cursor over each such cell to view the notes.)

We will follow these steps to construct the spreadsheet:

1. Lay out the time-ordered categories across the worksheet from left to right starting at cell A3.
2. In cell A6 (under the "Alternatives" label), type in the strategy titles from the strategy table (TIRE.XLS tab [Strategy Alternatives]). Type all the decision, uncertainty and value titles into row 3.
3. Next, ensure that all the nodes in the influence diagram are included in roughly the same order in which they appeared there.
4. Observe whether any computations need input nodes other than

those in the strategy table ([Strategy Alternatives] tab) and decision tree ([Schematic Tree] tab). If so, insert their titles between the existing nodes. It is generally best to add these new nodes to the decision tree, influence diagram and strategy table. In this case, we have added a category called "Present Value," however, because this is only a convenience for calculating "Net Present Value," we have not added this node to the other diagrams.

5. In row 6 and below, in the appropriate column, type in the possible values or alternatives that each category might take. If there is specific information for input nodes from the influence diagram, type it also in the respective columns. This will allow you to keep track of the input information for each node during the spreadsheet analysis.
6. Setting aside "Net Present Value" for the moment, start from the right-most node in Row 4, "Present Value". Type in the equation that will calculate Present Value using information from the cells to the left or below. Let this calculation take place in steps, that is, let the Present Value equation use the variables only as a first order calculation. There should not be a lot of parenthetical calculations, second order calculations, or combining of other nodes in the Present Value node. This principal, which should be followed throughout the other cell equation constructions, makes model changes and additions easier.
7. Take the next computational node on the left that Present Value calls for in its equation and type in the equation that will calculate it.
8. Keep doing this until you have related all the influence diagram nodes in row 4 on the worksheet back to the left except for the strategy. As you might see through equation generation, you are beginning to lay out the influence diagram in the equation-defined relationships for what is entered in row 6 and below of the worksheet.

Note that in this model, all the input cells are highlighted in yellow and that the equations are written to use these inputs in computations in either value or chance nodes or as alternatives in decision nodes including the strategy node. This allows rapid calculation of the present value of different alternatives with different value or chance combinations.

11.5 □ USING THE DECISION MODEL TO DETERMINE THE BEST STRATEGY

The decision model uses the inputs and relationships that are established through the decision modeling process to calculate a present value for each strategy.

In any analysis, there is a base case or "status quo" strategy, i.e., the result of no change from the present approach. In this example, the base case is the "Run" strategy. This means a bad tire condition with an associated probability of blowout of 0.7. The consequential cost of a blowout, weighted by the chance of occurrence during city and highway driving, provided the total expected consequential cost of blowout. In this case, the same expected consequential cost, $264, applies to either strategy decision. The risk of blowout for the base case is the probability of blow out (0.7) times the consequential cost of a blowout resulting in a $184 Risk (or expected value) of Blowout. Because this strategy involves no other costs, $184 is also the Present Value.

The alternative case is the "Replace" strategy. This means a good tire condition with an associated probability of blowout of 0.1. Applying the previously-calculated consequential cost of blowout produces a $26 Risk of Blowout. That plus the $50 cost of tire replacement results in a $76 Present Value for this strategy. Both of these results are shown in the present value table to the right of the present value result node.

The net present value for the "Replace" strategy is the Present value for the run strategy minus the Present value for the replace case. This is shown to the right of the present value table in a table for Net Present value and the positive net present value of $108 is also shown in the green-bordered cell. This is saying that the net is the value of avoiding the consequences of the base case by pursuing the cost and consequences of the alternate case. If, as in this example, the net is positive, then the alternative is the better decision. If there are several alternatives, the alternative that produces the highest Net Present Value is, in general, the best. The Decision Model provides a rapid calculation method for the present value of each decision strategy.

11.6 □ INSPECTION EXAMPLE

This section "walks through" an inspection strategy development for superheater or reheater tubes in a boiler. This example considers only economic factors. For safety-related components, the optimized schedule that you developed in the previous section should have been safety-constrained. In this section, you are looking for the most cost-effective way to determine whether the component conditions meet or exceed your predictions. If you want a "belt and suspenders" approach that uses safety constraints in the inspection strategy development, use the optimizer in Section 10 as a pattern to modify the procedures in this section.

11.6.1 □ Developing a Strategy Table

Now we will apply the decision model development process to a more realistic inspection program development example. We will develop an inspection program for high energy boiler tubes, such as those in a superheater or reheater. The damage mechanism is lost tube thickness caused by boiler slag erosion-corrosion.

We will consider thickness measurements with a "D Meter" or inspection with a 1/4 in. diameter ("small") longitudinal wave ultrasonic transducer with a digital thickness readout. We will consider these choices along with choices that involve different numbers of test locations, who will conduct the inspection, and how the inspection will be conducted. The example is contained in the example spreadsheet TUBESTRA.XLS, which is located on the Handbook CD-ROM. Copy the spreadsheet to your working directory and play along.

11.6.2 □ Defining the Candidate Strategies and Damage Mechanisms

First, we lay out the candidate strategies. In this case they are:

1. No Inspection (base case)
2. D Meter Thickness Meter
3. Small Transducer Thickness Meter.

Next, either from experience or from the literature, we list the damage mechanisms that apply to this particular component. (Remember, the inspection is about detecting and measuring the extent of the damage mechanism in the component.) List the damage mechanism(s) in the second column of the strategy table. In this example, the strongest mechanism consideration is stress rupture, and that is the only mechanism that we will consider.

11.6.3 □ Laying Out Inspection Strategy Categories

The first step in laying out the contents of the decision is to identify the categories of the decision model, from the failure probability information all the way to the decision criteria. Click on the tab, [Original Worksheet]. This template is shown in Figure 11.6.1. The column headings across the top of the table are for the categories of the information that are involved in the decision. Down the left side of the table is space for row headings that will contain the different candidate inspection strategies that will be compared.

Choosing the ultimate strategy is what the decision analysis is all about. Think of a category as:

- A quantity that you need to know before you can estimate the economic decision criteria or
- A resource that you will need to choose before you can estimate the economic decision criteria or
- A value that you need to calculate.

You can also think of these categories as things that you need to know or to do to perform an inspection. These categories are the items that will unfold after the decision is made and all which will ultimately determine the magnitude of the economic criteria.

You can list on a scratch piece of paper or "brainstorm" a list of the categories of information that you will need or the magnitude of which you will need to consider when you make the decision. A category list for a high energy boiler tube inspection might be:

- Net Present Value
- Probability of Failure Today Based on Trend

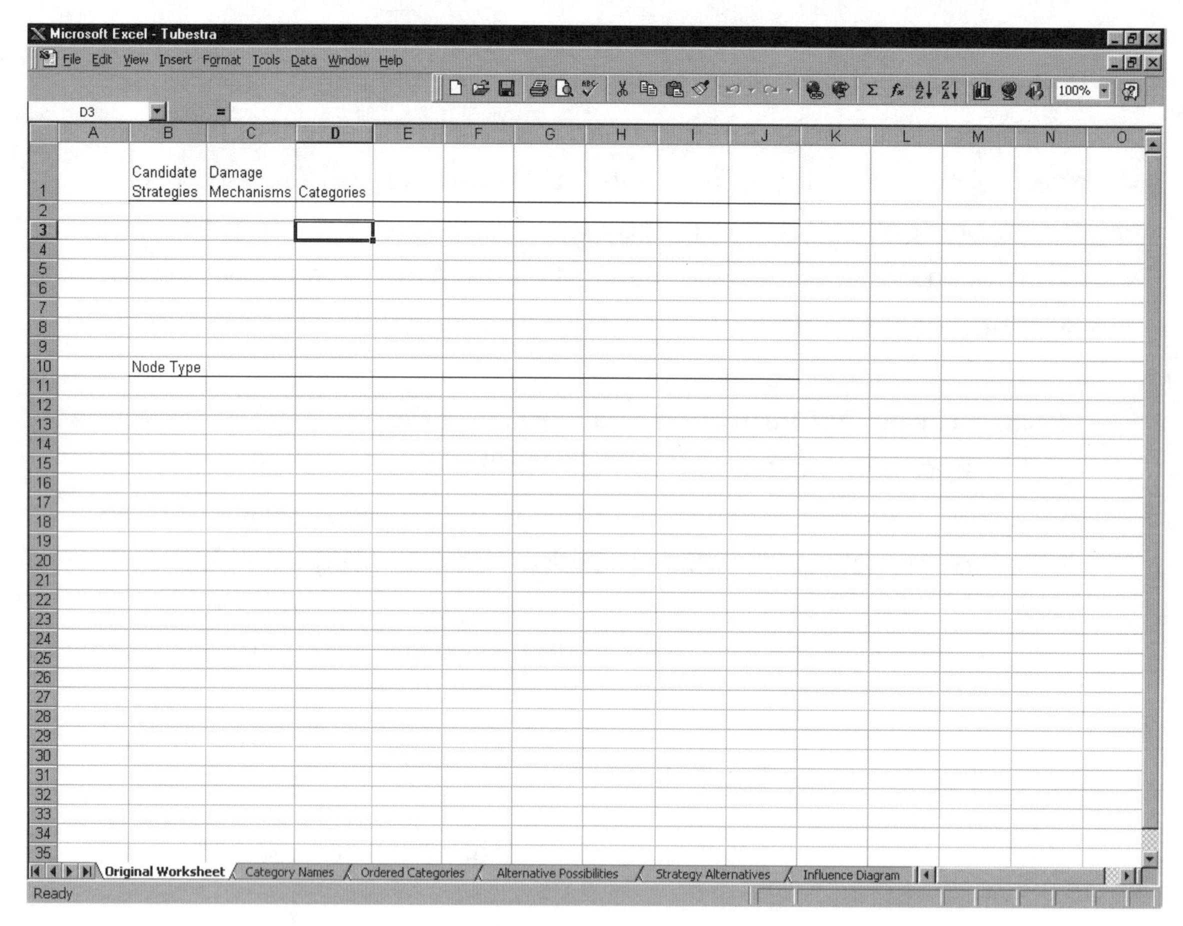

Figure 11.6.1 [Original Worksheet] in TUBESTRA.XLS

- Probability of Failure Given no Defect Reported
- Number of Tube Elements
- Inspection sensitivity
- Number of Elevations to Inspect
- Access to the component
- In-house or Contract Inspectors
- Consequential Time for Shutdown
- Opportunity Lost Cost per Hour

Shown in under tab [Category Names], these categories are listed from column three extending to the right, starting with Net Present Value. The rest are just listed in no particular order. Under tab [Ordered Categories] these categories have been ordered from left to right in the sequence in which they would expect to unfold in time.

You have now defined the constituents of the decision. The next task will be to "flesh out" these constituents with alternatives to be considered and with quantitative information.

11.6.4 □ Defining Decision Node Types

Look at each category and decide which type of decision model node that category represents. You have three choices:

1. Decision nodes are choices between two or several alternatives that address the mechanism(s) that you have selected.
2. Chance Nodes show the possibility of several outcomes in a category. Think about how each of these outcomes will have a certain chance of occurring. You will assess these probabilities later.
3. Value Nodes are single-value input constituents or calculations in the information flow process.

Under tab [Ordered Categories], the node types have been defined for each category for the boiler tube example. A hint on how to do this: Look at the type of information that you would put into each category. For example, tab [Ordered Categories] shows a category "Probability of Failure Today Based on Trend." This category needs a single numeric estimation from past data, so this node is a value node. On the other hand, the "Inspection sensitivity" category requires uncertainty information, so this node is a chance node.

11.6.5 □ Providing Alternative Choices

Look at each category, one at a time, in tab [Alternative Possibilities]. List the possible alternative choices for each of the categories and each of the candidate strategies. A hint here is to look at the node type that you have assigned. A decision node, as is shown under "Candidate Strategies," requires specific choices. As you can see, the specific choices here are "No Inspection," "D Meter Thickness Meter" and "Small Transducer Thickness Meter." However, for a chance node, such as "Inspection Sensitivity," the possible outcomes are shown to be "50% of Actual," "100% of Actual" and "150% of Actual" for the D Meter; and "90% of Actual,"

"100% of Actual" and "110% of Actual" for the Small Transducer. The decision model is beginning to take shape.

11.6.6 □ Linking the Alternative Choices

Now we start linking things together to define the different inspection strategies.

1. Look at tab [Strategy Alternatives]. The candidate strategies on the left make a natural starting point. The first candidate strategy is circled in green to help you focus on it.
2. For each candidate strategy, circle the alternative choice in the next category to the right that is associated with the inspection for that failure mechanism. Remember that the alternatives and possibilities in the categories are grouped by the several alternatives or possibilities as you wrote them down in the previous step. When you are doing the circling, remember to group matching alternatives or possibilities within the same circle.
3. Continue this process for each node that applies, as you move from left to right across the categories.
4. Go back and link the circled alternatives and possibilities with straight lines that join adjacent categories across the table. Performing the same process with red to develop the D Meter strategy and with blue to develop the Small Transducer strategy exhausts the possibilities.

The finished product will look like tab [Strategy Alternatives].

11.6.7 □ Developing an Influence Diagram

We will now construct the influence diagram for the first-circled strategy. Probably the easiest way to do this is to take each element from the strategy table and place it in time order from left to right as unconnected nodes. Start with the value node on the far right and ask, "What do I need to know to calculate NPV?" If the existing node to the left contains all the information that is directly needed, then connect the nodes with an arrow. If something intermediate is needed to relate the existing node to the left to the NPV node, then add another node. Note that the value nodes are green-bordered rectangles, the chance nodes are ellipses and the decision nodes are red-bordered rectangles.

Work to the left, taking each node and asking the question "What do I need to know to calculate this node?" We then connect existing nodes or add new ones on the left and connect them with arrows until we have connected all the way back to a strategy alternative decision or have defined nodes for which the information will have to be determined directly. The finished diagram for the first circled strategy is shown in tab [Influence Diagram]. Take a moment to follow the right-to-left construction in this example.

11.6.8 □ Developing a Decision Tree

All the decision and chance categories in the strategy table are to be represented in the decision tree. The best way to accomplish this is to take the time order of the categories in the strategy table. From tab [Strategy Alternatives], the strategy table has been ordered to reflect the time-order sequence of the categories from the candidate strategies. The following steps have been performed to develop tab [Schematic Tree].

1. On the left is the strategy decision. Note that the three strategies are listed in the blue background area to start the tree. They are spaced to allow the other categories in the tree to be developed to the right.
2. Looking at the diagram generally, to the right are the consequences that result from the inspection strategy decision. We show these in red. The heavy black line to the left of these consequences shows that each consequence applies only to the respective decision, i. e., strategy. Each black line extends vertically the height of the decision(s) to which it applies.
3. Looking at the diagram specifically, to the right are the decision choices for the two inspection strategies. Note that the decision choices are the same for both the strategies. This is indicated by the black line vertically the height of both inspection strategies. The black line indicates that the decision choices apply equally to both decision strategies regardless of the difference in inspection sensitivity. These choices have a blue background because they are the choices in a decision.

4. To the right are the two other decision nodes. They are treated the same way because they are independent of the inspection strategy.
5. To the right, finally, is a chance consequence that results from the uncertainty of time the unit might be shutdown if a tube failure occurred. Again, these uncertainties are independent of the inspection strategy, so the vertical height of the black line ties all the inspection strategies together, including No Inspection, because it would also apply to that strategy.

Do the above until all decision and chance nodes are included in their proper time order.

11.6.9 □ Concept of Spreadsheet Development

The spreadsheet decision model is constructed by laying out the spreadsheet cells from left to right in the same time order as was used in the strategy table. The equations within each cell will represent the relationships that are described in the influence diagram. The multiple input values for each cell in the decision model are described in the schematic decision tree. The spreadsheet model must contain the elements of:

1. The decisions for each strategy
2. The uncertainties within each strategy
3. The subsequent decision within each strategy
4. The computations that relate the decisions and uncertainty to value.

To construct the spreadsheet, we lay out the previously-established categories from left to right across the top of the spreadsheet as they were in the strategy table. The supporting tables for calculations and data input will be placed below their respective categories. Note the decision model layout in tab [Decision Model]. Notice how the column headings indicate the order of the basic categories across the top of the spreadsheet as in the strategy tables.

11.6.10 □ Developing the Spreadsheet

Spreadsheet construction will follow these steps:

1. Lay out the time-ordered categories, by their titles, across the spreadsheet from left to right starting at cell A3.
2. Under the Alternative label, type in the strategy titles from the tab [Strategy Alternatives]. Then type in all the decision, uncertainty and value titles. For the model calculation purposes, the strategy to be calculated is entered in the yellow background cell, A4. (If you hold the cursor over the colored background nodes, annotations will be displayed that indicate the cell input.)
3. Next, ensure that all the nodes in the influence diagram are included and in roughly the same order.
4. Examine the [Strategy Alternative], [Influence Diagram] and [Schematic Tree] tabs for each strategy and observe whether any input nodes have been added to support the computations. If so, insert their titles between the node titles from the decision tree where the influence diagram indicates that they fall. It is best to insert these same nodes in the schematic tree, influence diagram and strategy table.
5. Across the [Decision Model] worksheet you will find that you have, across row 3, the decision tree time order layout with the added computational information from the influence diagram.
6. In row 6 and below, in the column that corresponds to the category or node, type in the possible values or alternatives that each category in the strategy table might take. If there is specific information for input nodes from the influence diagram, type it also in the respective columns. This will allow you to keep track of the input information for each node during the spreadsheet analysis. The yellow background cells indicate input cells for the decision model to use in the calculations.
7. Starting from the node on the far right in Row 4, "Present Value," type in the equation that will calculate "Present Value" using information from the cells on the left. Let this calculation take place in steps, that is, let the "Present Value" equation use the vari-

able(s) only as a first order calculation. There should not be a lot of parenthetical calculation, second order calculations, or combining of other nodes in the "Present Value" node. Follow this principal throughout the cell equation constructions. This makes model changes and additions easier.

8. Take the next computational node on the left that "Present Value" requires in its equation and type in the equation that will calculate it.
9. Keep doing this until you have related all the influence diagram nodes in row 4 on the worksheet back to the left except for the strategy. As you might see through equation generation, you are beginning to lay out the influence diagram in the equational relationships for what is entered in row 6 and below of the worksheet.

11.6.11 □ Determining the Best Strategy

To determine the best strategy, insert the values for each strategy in each of the categories that require input in the [Decision Model] tab and note the "Present Value" that the model calculates in the green background cell. The base case or "status quo" strategy is the one that would occur if there were no changes from the present approach. This would be the No Inspection strategy in this example. This strategy has an associated full exposure to the estimated 10% probability of tube failure today.

The next calculations would be the consequential time of shutdown distribution. The subsequent calculation of the expected consequential cost of the base case is the product of the number of tube elements, the present probability of tube failure, the expected value of the consequential time of shutdown distribution and the opportunity lost cost of an hour of shutdown. The Present Value is the expected consequential cost plus the inspection cost, which in this case is zero, or $147,000.

The alternative cases are the other combinations. They use this same calculation scheme to establish the expected consequential cost by the weighting the effect of the inspection sensitivity distribution on today's probability of failure. Of course, in this case the respective inspection strategy cost is added to the expected consequential cost to provide the Present Value for the respective strategy and decision combinations.

The subsequent decisions in these cases determine personnel to use, the number of elevations to inspect and the method of access to the High Energy Boiler Tubes. As shown to the right, the "Net" is created by subtracting the alternative "D Meter" or "Small transducer" case from the base, "No Inspection," case. This is because the net difference in the decision is the net value of pursuing the alternative case over the base case. This is saying the net is the value of avoiding the consequences of the base case by paying for the alternative case.

If, as in this case, the net is positive, then the alternative is a better net present value producing decision then the base case strategy. The increases in value obtained by pursuing the alternative cases over the base case are shown in the Net Present Value table on the far right for all the possible strategy and decision combinations. The differences come from the table of Present Value Combination calculations just to the left of the Net Present Value Combinations.

The largest and most favorable Net Present Value is for the Small Transducer Thickness Meter strategy together with using in-house resources to do the inspection at 4 elevations with a sky climber. The result of $113,720 is shown in cell AD12 in a green bordered cell. Note that the "Decision Model" role is to provide a rapid method of calculating the present value of each decision strategy combination to determine the greatest value-producing combination.

NOTES:

SECTION 12

To perform this procedure, you will need:

A properly functioning decision model. For this Handbook, such models are developed in Sections 10 and 11.

This procedure will produce:

- *Greater depth of understanding about how the inputs to your decision model affect the decision as you have modeled it.*
- *The identity of the variables about which more information is most valuable.*
- *An estimate of the possible value of additional information.*
- *Ideas for improving information that is important.*

The next step will, in general, be a return to the procedure that brought you here. If, after performing this procedure, you are satisfied with your model, you will go to the procedure that follows the referring procedure.

SENSITIVITY ANALYSIS 12

A major question that often needs to be answered is whether the information in each input node in a decision model is sufficiently well-defined to support the analysis. How well-defined the information at a node needs to be depends upon how much variation in that node's information influences the NPV. The degree of influence is the sensitivity of the decision model to the input node. We need to determine which model nodes have the highest sensitivity and then ensure that we have sufficiently well-defined information for these nodes. This is the goal of a sensitivity analysis.

Extremely important note: This procedure can help you to make a good model better. Although it might (with extreme good luck) point you to a flaw or weakness in your model, you cannot rely on this procedure to identify missing inputs, failed logic, typos or other modeling mistakes. Computer people have an expression that has become a cliché: garbage in, garbage out. You must understand that this procedure can only tell you at which inputs garbage (bad data) will hurt you most. It cannot tell you whether your data is garbage, it cannot directly tell you how to make good data out of garbage if that is what you have, and it most certainly cannot tell you whether your model is making garbage out of good data.

With these limitations in mind, sensitivity analysis is always a good idea in any risk-based work. Once you build a decision model, you will probably exercise it many times over the life of the components that it addresses. Sensitivity analysis is a prime quality control tool. Sensitivity analysis should be considered absolutely necessary if a model produces results that disagree with your intuition or with expert opinion.

12.1 □ EXERCISING A SPREADSHEET TO IDENTIFY DECISION-CONTROLLING VARIABLES

We will explain sensitivity analysis by exercising the spreadsheet model that we built in Section 11.6. Proceed as follows:

1. Make sure that @Risk is installed and that it loads when you start Excel by using the @Risk icon or (in Windows 95) the Start Menu Program item. See the @Risk manual if you have any problems.
2. Load the example spreadsheet TUBSTRSK.XLS from the Handbook CD-ROM and copy it to your working directory. This is a version of the TUBESTA.XLS spreadsheet model from Section 11.6 in which the equations have been modified to allow for sensitivity analysis for the uncertainty categories of "Inspection Sensitivity Effect" and "Consequential Time for Shutdown."
3. Click on each of the uncertainty nodes in the spreadsheet in cells D4 and J4. Note the @Risk functions that provide the uncertainties in the equations. In this case, we used the discrete distribution function, RiskDiscrete, rather than a more complicated continuous distribution like the normal. In a first cut decision strategy analysis, it is more efficient to use an approximate distribution like a discrete distribution until we locate the highest decision sensitivities and determine whether the values that are produced by the different strategies are even close to each other.
4. Click on the cell M4, Present Value, and click on the icon "Add the selected cells as @Risk Output." This will allow us to see how sensitive Present Value is to the two uncertainty cells.
5. Set the model at 100 iterations for this analysis.
6. The strategy cell, A4, shows "D Meter" to set the model up for an evaluation of this strategy.
7. Click on the icon "Run Simulation" to start the Monte Carlo analysis. When the analysis is complete, the @Risk results screen and the simulation statistics screen appear.
8. In order to see the most sensitive random variable, click on [Sensitivity] on the Results screen. Click on the [Graph] icon, select the check box [by correlation] and click [OK]. The graph appears when it is maximized. You now have the indication that the most sensitive variable for the D Meter at one inspection level with a sky climber with a contract inspector is the "Consequential Time for Shutdown."
9. You can continue to perform these sensitivity analyses by choosing different values for the decision variables and rerunning the sensi-

tivity analysis. You will be looking for the random variables that consistently dominate the value of present value.

There are usually just two or three input variables that control the output of the decision model.

12.2 □ METHODS FOR REFINING SIGNIFICANT VARIABLES

After you have used the following methods to refine the inputs, use the process in Section 11.5 or 11.6.11 to insert the new information into the model and to determine the best strategy.

12.2.1 □ Multiple Source Plant Maintenance Personnel Interviews

If the source for one of the more sensitive variables is a "rough estimate," then it would be best to go to the responsible manager and seek a better information source. If the manager agrees with the idea of an interview, then conduct one (See Section 9.3.1) and enter the information directly into the appropriate random variable distribution. If there are two people that are considered to be good sources, then you can simultaneously interview both of them. As before, seek consensus between the subjects during the coin-stacking process.

12.2.2 □ Refined RLA Models

If the sensitivity analysis shows that one of the most sensitive inputs is the failure probability in the column for damage mechanism, then you might need to refine this input. If just interviews and failure history were used alone or were combined, then a detailed RLA analysis might be needed. If you perform a new, more detailed and rigorous RLA analysis, then you may want to use its failure probability versus time information alone without using a Bayesian-like combination. See Section 9.4.

12.2.3 □ Assessing the Value of a Significant Variable

If one of the most sensitive variables is one that you can control at a different value, such as "Number of Elevations to Inspect," then move its inputs to high and low while you leave all the other values at their mid range or nominal value. Observe the difference in present value that you calculate for each value. If the present value difference that you create is greater than the incremental cost difference between the lower value and the higher value, then you have a basis for using the higher value of the variable. For example, with the "Small Transducer," the present value that is created by inspecting at 4 elevations instead of at 1 elevation is much higher than the cost of the increased time that is required to inspect more locations. This means that inspecting at 4 elevations is cost effective. This illustrates a "positive value of information."

If, as in Section 12.1 step 8, the most sensitive variable is an externally calculated value, you have several options. You could "sharpen your pencil" and more precisely calculate the value. Or, you could calculate "best case" and "worst case" values and then calculate the NPV spread between the two. Or, and this arguably is the best option (after all, this is a probabilistic environment!) you could make the sensitive variable(s) @Risk inputs by assigning distribution(s) and running a simulation to obtain a probability distribution for NPV. And that, should you choose to accept it, is our parting challenge; your homework assignment if you prefer. It will take you into the next level of this technology.

NOTES:

NOTES:

SECTION 13

DEFINITIONS

DEFINITIONS 13

The following definitions are tailored to the Handbook. Rather than being stiffly formal, they are designed to give functional understanding of terms used in the Handbook. If you need more formal definitions, consult the Fault Tree Handbook (Appendix C.1), one of the references (Section 15) or the other volumes in the risk-based inspection program development series.

Best Decision—The decision that the presently available information suggests will produce the highest value of the decision value criteria. Obviously, the best decision can only be selected from the options that can presently be identified. Further, there is no guarantee that the best decision now will prove to be the correct decision at some later time. This is a result of circumstances and knowledge changing with time. Also more events do occur;

Components—The basic units of systems. You can get seriously hung up on terminology if you're not careful; using terms like "element," "part" or whatever can give you a communications nightmare. One model's component can be another model's element, etc.. Therefore this Handbook uses the term "component" as the lowest-level "building block." Obviously, a feed pump could be the system in one analysis, a subsystem in another and a component in a third; if you're consistent within an analysis and within a given database, you should stay out of trouble.

Confidence—As the name implies, this is the "feel good" factor. Viscerally, if you overhaul your turbine and find no serious problems, you may, after some considerable analysis, conclude that you can confidently run a lower overhaul frequency. Risk-based methods provide a good tool for quantifying this confidence. Sections 9.4.3 and 9.4.4 show you how to use inspection results to "update" your knowledge base. Note, however, that the confidence to which we refer is *confidence in the data*. The data in which you become more confident might point toward a high probability of failure. Although there is considerable probability theory that involves confidence and "confidence intervals," this Handbook does not apply that much formality. See the Fault Tree Handbook, Appendix C.1, for more information.

Consequence—The impact (results) of a failure, for example: forced outage or derate, structural or other damage, fire, explosion, injuries, deaths or environmental impacts. Obviously, the consequences listed above could be measured in terms such as, respectively, full-power hours or equivalent full power hours, number of ruptured waterwall tubes, x% loss over y ft^2, x psi overpressure over y ft radius from point of origin, lost hours, number of deaths, or x pounds of y substance released over z area. None of these measures, however, are generally useful, because they do not "talk" to each other. Risk-based programs therefore generally use a financial consequence measure, dollars, or limit the consequences in other ways. For more discussion, see the executive summary and Appendix A.

Cut Set—A particular combination of specific failures caused by specific failure modes, which yields a **failure consequence** that leads to an undesired consequence. Generally, there are many cut sets associated with each failure consequence because a variety of failure combinations can cause the same loss of system function. A minimum cut set is one that has the minimum number of failures necessary to lead to the undesired consequence. Generally, only minimum cut sets are analyzed, and the word minimum is often not stated.

Decision Analysis—A formal process for organizing the information, including measures of information uncertainty, that will influence a future course of action or inaction, mathematically modeling the information, and making calculations that will help identify the **best decision** and/or determine what additional information is necessary.

Deterministic—the nature of traditional engineering analysis. Deterministic analysis uses point estimates or actual measurements in calculations. Deterministic methods are inherently conservative because, for safety reasons, they must generally use "worst case" estimates for any values that cannot be measured. For example, maximum flaw size generally cannot be measured in even a new component; therefore a deterministic analysis must assume the largest possible flaw exists at the worst location. Because the "largest possible flaw" is itself **uncertain**, a factor of safety is generally needed on that as well.

Event Tree (Analysis)—an analytical tool that organizes and characterizes potential failures in a methodical manner. The analyst identifies potential initiating events, then for each initiating event, identifies a logical chain of failures, mitigating events and propagating events. The chain branches at each event into two or more paths, with each path leading to another event or a consequence. If probabilities for each branch are known, and if values can be calculated for each consequence, the risk associated with each path through the tree can be determined.

Failure Sequence—an ordered sequence of events that leads from an initiating event to an undesired consequence. The sequential events may be failures, mitigating events or propagating events. Failure sequences are often analyzed with and their probabilities calculated using **event tree analysis**. Dominant failure sequences are those associated with the highest risk.

Failure Severity—the performance of a component when it is taken out of service for repair or replacement. Three severities are generally identified:

1. Incipient means the component is still functioning and providing the required services at normal parameters but has a condition that makes its immediate removal from service necessary (e.g., a pump with elevated bearing temperature).
2. Degraded means the component is still providing some service but at reduced performance (e.g., reduced flow from a pump with damaged wearing rings).
3. Catastrophic means the component is not performing its function (e.g., a valve that doesn't close or a pump that doesn't provide any flow).

Fault Tree (Analysis)—A fault tree is a system model that shows the logical relationship between the system components and how failures of these components can cause system failure. Fault tree analysis deduces these interrelationships in a "top-down" fashion. The analyst specifies the system failed state, then determines subsystem failure combinations that will cause system failure, then progresses stepwise into more and more detailed specifications of what can cause subsystem failure, until the failures identified are of basic components which need no further subdivision. A fault tree is a qualitative model of the system until probabilities are assigned to the component failures involved (the tree is "quantified"); at which time the model becomes a quantitative model from which system failure probability can be calculated. See the Fault Tree Handbook, Appendix C.1, for more information.

Human Error—Any human action or omission which deviates from established or generally recognized acceptable procedures, and which could result in an undesirable event.

Human Factors—the field of study that deals with the "man-machine interface." Although this Handbook and its examples deal almost exclusively with structural and operational issues, **human error** can obviously be a critical factor. This is particularly true in fossil fuel-fired power plants and other industrial facilities because automatic protection systems and redundant systems are relatively rare. Paradoxically, much of the available human factors research originated in the nuclear industry, which generally does have such systems. Although integrating human factors analysis is beyond the scope of the Handbook, we have included an introduction to the currently available methodology in Appendix D.

Inspections—activities that gather information and thereby reduce **uncertainty**. Tests are special kinds of inspections that determine whether a component or system is capable of functioning. Inspections in general improve our knowledge about the current state of what is being inspected. "Inspection effectiveness," "probability of detection," time out of service to perform the inspection and probability of inspection- or test-induced damage, in addition to the issues concerning the test or inspection itself, need to be considered during risk-based program development.

Minimal Cut Set—See **Cut Set**.

Modeled—modeling is the process of constructing a mathematical tool or series of equations that will "act like" the structure or event being modeled. For example, a finite element analysis models a complex structure with a series of small "pieces" whose individual behavior is simple enough to be calculated precisely and predicts stress levels in various parts of the whole structure. Risk-based methods use a **probabilistic** model of a system or process to analyze the system or process and predict its future behavior and the uncertainty in that behavior. Model users must satisfy themselves that the models they use are valid. Risk-based program models in this Handbook are generally a logical departure from historical fact. Formal model validation is beyond our scope.

Net present value (NPV)—the present value is the value contributed by an asset in the future stated in equivalent present-day dollars. The net present value in this handbook will generally be the difference in present value between what your facility would normally do in the future and what you are proposing to do. When calculating an NPV, you will appropriately discount future after-tax incomes and expenses. All the examples in the Handbook use NPV as the value measure, however, you don't need to become an accountant to use them, at least not at first. In risk-based optimization, you might never include any revenue streams in your calculations; if not, your best strategy will be that which produces the most positive or least negative NPV savings. For more information, dust off (if necessary) and open your Engineering Economy textbook, or consult one of the references in Section 15.

Present Net Worth—See **Net Present Value**.

Probabilistic—the nature of most work in risk-based methods. The goal is to use all the information available, even if it is uncertain. If a parameter can reasonably assume different values, then all of those values are used in accordance with the expectation that they will occur. Probabilistic methods often use distributions that are centered on the "best guess" for values which cannot be precisely measured. Any predicted values for important parameters such as component flaw sizes in future years must by definition be treated probabilistically because they cannot be known until they occur. The only alternative is a **deterministic** approach that is unworkable because it must use excessively conservative tools such as "worst case" assumptions.

Probability—a measure of likelihood. Probability is a number between zero (impossibility) and one (certainty). Probability looks very much like frequency and, in risk-based calculations it behaves very much like frequency, but it is more than that. For starters, if you think of it as "average" frequency or "instantaneous average" frequency, depending upon whether the specific application specifies a time period, you will not be far off. See the *Fault Tree Handbook*, Appendix C.1, for a formal definition and more information.

Probability Distributions—provide a measure of the most likely value of a parameter and the **uncertainty** of that measure. If a parameter can have a range of values any of which are equally likely, the distribution is said to be uniform. In most cases, the distribution will have a "peak" at its most likely value. The procedures and examples in this Handbook generally use Wiebull distributions to approximate "real" distributions, for reasons that are discussed in the sections that introduce quantitative risk assessment. The "Big Idea" is to unanchor yourself from thinking **deterministically** and recognize that all estimates and most measurements involve some amount of **uncertainty**.

Qualitative Risk Analysis—A risk analysis that can rank a limited range of **components** or **systems** according to the risk they pose, but which cannot measure the risk other than by general terms such as "high," "medium" or "low." The advantages and limitations are discussed more fully in Section 5.

Quantitative Risk Analysis—A risk analysis that:

(1) Identifies and delineates the combinations of events that, if they occur, will lead to an undesired event
(2) Measures or estimates the frequency of occurrence for each combination
(3) Measures or estimates the consequences
(4) Calculates the risk associated with each combination; and
(5) Uses the calculated risk values to help develop and select **strategies** or make other decisions.

The analysis can also integrate relevant information about facility design, operating history, component history, and human actions into a single framework or model which, when evaluated probabilistically, provide both qualitative and quantitative insights about risk and risk control.

Remaining Life Analysis (RLA)—An engineering tool that mathematically combines a damage mechanistic model with service-related inputs to model damage progression over time. The model "detects" failure when the damage exceeds that which the component can tolerate, e.g., when previously determined failure criteria are met. The methods themselves are briefly described in Section 9. Consult the references (Section 15) for more information.

Remaining life assessment (RLA)—See Remaining Life Analysis.

Risk—a two-dimensional measure of adversity. It combines probability and consequence. For example, crossing a street exposes you to the risk of being struck by a vehicle. Note that the risk is as inextricably associated with the activity as is mass with a quantity of matter. You can control the risk by, for example, looking both ways before you cross, but you cannot eliminate the risk except by eliminating the activity. Risk is mathematically the product of event probability and event consequence. If the probability of a tube failure is 10^{-2}/yr and the consequence of the failure is $200,000, then the risk is $2,000/yr. Think about risk like power in a circuit. If you know the power in the circuit, you know things that neither the voltage nor current alone can tell you, however, in some circumstances, you need to know voltage and current as well. Knowing the current would be nice when you're sizing the wire, for example. Likewise, although risk is an extremely useful parameter, you must also track its components. An event of little consequence can be unacceptable if it happens too often, and a rare event can be unacceptable if its consequences are excessive. The reason for this apparent irregularity in the risk function is risk tolerance. Risk tolerance can be measurable but it is often psychological, seldom linear and difficult to model. The financial world knows the product of probability and consequence as "expected value."

Risk Importance—a measure of relative risk. Many risk importances have been defined, but most belong to one of two categories. The first category addresses the absolute or fractional risk to which failures or a given component (or system) contribute. The second category addresses the conditional risk associated with assumed failure of the component. Consider two components, each of which contributes 10% to the total risk associated with an operation. Measures in the first category would assign each the same risk importance. However, if the failure probability of one component is low and the consequence high, while the opposite is true for the other component, the second category of risk importance measure would assign different risk importances. This is because the component with the high associated consequences, given that the measure assumes it will fail, has a higher risk importance. The Fault Tree Handbook, Appendix C.1, further discusses risk importance.

Safety Limit—A constraint placed on the probability of injury or fatality from an equipment component failure. This probability usually contains the probability of the failure and the probability that the failure would cause an injury or fatality. For simplicity, this handbook assumes that a failure will cause a fatality if the component is classified as safety related. This may, of course, be a very conservative assumption.

Sensitivity Analysis—a technique used in mathematical modeling that determines which inputs are the most important contributors to the value of the output. Generally, a sensitivity analysis "perturbs" each input, measures the effect on the output, then ranks each input according to its affect. In probabilistic risk assessment, we use sensitivity analysis to find out where resources can best be spent to reduce **uncertainty**.

Strategy—a defined course of action. All significant factors need to be considered in the definition. For example, defining a strategy for something apparently simple, like "liquid penetrant inspection of sootblower bracket attachment welds" could require, in addition to all the variables associated with the inspection itself, the interval or frequency, who will do the inspection, whether scaffolding or a powered platform will be used, etc.. When you use risk-based methods to design an inspection strategy, you must identify all these variables and account for them in your **model**.

Subsystems—units of **systems**. Depending upon the scope and complexity of your analysis, you may or may not want to identify subsystems. See also **components**.

Systems—units of the facility that will be the subject of analysis. In the Handbook, we call the "whole thing" that is being analyzed the system, no matter what it is. The system is broken down into **components**; you may or may not want or need to identify **subsystems** or **trains**. See also **components**.

Train—a system or subsystem that is duplicated. Sometimes, you can consider trains as identical, however, individual trains and identical components in different trains seldom have the same operating history or failure rate. You might initially assume like trains to be the same, but structure your model for later update when data about train "individuality" is gathered.

Unavailability—the probability that a component is not in the operating state or available for service. One definition:

$$\text{Unavailability} = \frac{\text{Down time}}{\text{Down time} + \text{Time between repairs}}$$

Other definitions are possible.

Uncertainty—a measure of the **confidence** with which a value is known. The uncertainty is graphically represented by the "width" and "flatness" of a distribution. The lower the uncertainty the "narrower" and "steeper" the distribution becomes. Note, however, that variability of the source, which may or may not be controllable, will also affect the shape of a distribution. If there is no uncertainty (or variability), the "distribution" collapses to a single point at the known value. This, however, is a most unusual occurrence.

SECTION 14

CLOSING COMMENTS

CLOSING COMMENTS 14

This Handbook covers a lot of ground. Although it contains nothing new, it packages bits and pieces of many classical disciplines in sometimes unusual ways. It will not make you an expert risk analyst, nor will it make you an expert in any of the underlying disciplines. Remember the stated purpose: to introduce you, an expert in your facility, to a set of tools that can help you. There are two key words in that sentence: introduction and expert. An introduction to a tool does not make you a skilled craftsperson. That requires considerable practice and possibly coaching by other craftspeople. And you must be an expert in your facility, or you will be unlikely to account for all the hazards that you face.

Although this handbook is based upon fossil fuel fired power generating station equipment, it is directly applicable to most other industrial equipment. However, the Handbook and its examples introduce techniques. The combined procedures do not explicitly address all the hazards and risks associated with a complete generating facility, nor even of any single piece of equipment within such a facility. Boilers, for example, have fuel systems. Fuel systems are mentioned as a possible outage source; they are also a possible fire source. Fire is no where addressed explicitly, but some fuel equipment failures could cause fires, which could cause really long outages and severe safety problems. When you design a risk-based application, you must consider all the risks and all the hazards that any system or component under analysis could present. You may, upon reflection, decide not to consider various hazards or risk elements in your analysis. Simplifying a problem logically can be good engineering; simplifying by ignorance is not.

In addition to all of the other cautions and warnings that were issued throughout this book, there are two technical limitations about which you should be aware.

The first involves the optimizer that is discussed in Section 10. The MANTOP^{TM} optimizer that is offered in Section 10 is not robust because it uses SOLVER in EXCEL. The SOLVER routine is relatively unsophisticated and, as applied by MANTOP^{TM}, will "choke" on more than four "projects." Other issues were noted in Section 10. If you need to optimize a larger number of inspection programs or if the optimizer in this handbook does not provide a satisfactory solution, you might want to substitute a more robust optimizer. See Appendix F if you need help.

The other involves the remaining life assessment (RLA) model for piping inspection with safety considerations that is discussed in Section 9.4.4. The PIPESAFE.XLA spreadsheet probabilistic model needs to be applied cautiously. It is not to be used for longitudinal seam weld indications in steam piping because of the complexity of the damage propagation mechanisms in this situation. The model is usable for a first cut prioritization that will be followed up with more detailed inspection and analysis for safety-related components. For more information or for assistance with RLA modeling, see Appendix F.

SECTION 15

REFERENCES 15

AIChE, *Guidelines for Hazard Evaluation Procedures with Worked Examples*, 2nd Edition, AIChE, New York, 1992.

American Society of Mechanical Engineers, *Risk-based Inspection—Development of Guidelines, Volume 1, General Document*, ASME Task Force on Risk-Based Inspection Guidelines, Washington, DC, 1991.

American Society of Mechanical Engineers, *Risk-based Inspection—Development of Guidelines, Volume 3, Fossil Fuel-Fired Electric Power Generating Station Applications*, ASME Task Force on Risk-Based Inspection Guidelines, Washington, DC, 1994.

ASM Metals Handbook, Vol. 10, Eighth Edition, Failure Analysis and Prevention, ASM International, Metals Park, Ohio, 1975.

ASTM Website, www.astm.org, American Society for Testing and Materials, West Conshohocken, PA.

Brealey, Richard A., and Myers, Stewart C., *Principles of Corporate Finance*, Fourth Edition, McGraw-Hill, New York, 1991.

Brigham, Eugene F., and Gapenski, Louis C., *Financial Management: Theory and Practice*, Sixth Edition, Dryden Press, New York, 1991.

Hogarth, Robin, *Judgement and Choice*, Second Edition, John Wiley & Sons, New York, 1987.

Schmitt, S. A., *Measuring Uncertainty: An Elementary Introduction to Bayesian Statistics*, Addison-Wesley, Menlo Park, CA, 1969.

Swain, A. D., *Accident Sequence Evaluation Program Human Reliability Analysis Procedure*, NUREG/CR-4772, US Nuclear Regulatory Commission, Washington, DC, February, 1987.

Swain, A. D., and Guttmann, H. E., *Handbook of Human Reliability Analysis with Emphasis on Nuclear Power Plant Applications*, NUREG/CR-1278, US Nuclear Regulatory Commission, Washington, DC, August, 1983.

Viswanathan, R., *Damage Mechanisms and Life Assessment of High Temperature Components*, ASM International, Metals Park, Ohio, 1989 , pp. 232-233.

APPENDIX
WHY FINANCIAL METHODS?

The Handbook is going to bring you into the world of finance. Some engineers have had no exposure to finance since the degree course in Engineering Economy. Many engineers have dealt with cost issues; however, few have dealt with financial issues. This appendix discusses why you need to start getting acquainted with such issues and use financial tools.

Several developments in recent years have substantially changed maintenance practices:

- Engineers started using analytical tools to justify maintenance decisions. The selected tools, however, were engineering-based, and the resulting analyses could not be fairly evaluated by financially-oriented corporate decision makers.
- Predictive maintenance techniques have developed and matured. The techniques, however, require substantial investment, also, they removed the predictability enjoyed by traditional time-based preventive methods. These methods also provide results that are only understood in an engineering context.
- New equipment is designed for longer operating intervals, but, because of newly-developed design tools, it can be built with less margin than was common in older equipment.
- Older equipment still in service is aging to the degree that accelerated deterioration, or "wear out" is possible. Because of competition, replacing much of this equipment cannot be justified.

All these factors argue for an analytical framework that will ensure safety and engineering reliability and that will present conclusions in a way that corporate decision makers, stockholders and regulators can understand.

Because corporate investment decisions are generally financial decisions, engineering analysis applied to maintenance needs to fit into the world of corporate finance. To gain access to corporate investment resources, engineering has to feed into an existing process. Decision analysis constructs a bridge between the worlds of engineering and finance. In addition, the financial decision process requires engineers to more rationally account for uncertainty than by using worst case analysis. The tool of choice is probabilistic analysis, or in the case of component life, probabilistic remaining life analysis. Note that financial decision makers already recognize the uncertainty in decision making and therefore use expected values.

When they use probabilistic analysis, engineers must not only estimate, for example, component life, but they must also state the spread or uncertainty of the estimate. The decision analysis model takes the engineering uncertainty and, by modeling the failure consequences, i.e. lost production cost, etc., determines the expected cost of the forecast failure. (The expected failure cost is the estimated failure cost at the time of expected failure, in present dollars, times the failure probability.)

Engineering analysis results that are expressed in the probabilistic format may be used in the world of corporate financial decision making. Predictive maintenance decisions can then compete on a level playing field with other corporate investment decisions.

In decision analysis, safety decisions can be handled by setting failure probability limits that override any financial criteria. This procedure fits many existing corporate "safety first" policies, and allows safety-related criteria to be considered simultaneously in the same decision model as finance without assigning a financial value to life.

APPENDIX B
COMPUTER AND SOFTWARE REQUIREMENTS

To use the Handbook and its supplied resources, you will need a personal computer with the following resources and capabilities. We will not attempt to specify precisely what sort of processor, how much RAM, how much disk space, etc.; what will work and what is practical in terms of operability and speed will be up to you.

- The Handbook templates or example spreadsheets and spreadsheet macros are written in Microsoft Excel 97 for Windows. Although you may be able to translate them into other versions of Excel or for use with other spreadsheet software, if you choose to follow this course of action, you're on your own. Before you attempt to use any of the spreadsheets, start Excel and ensure that "Analysis ToolPak," "Analysis ToolPak—VBA" and "Solver Add-in" are checked under [Tools], [Add-ins] menu. If any are missing, install them. Consult the Excel documentation if necessary.
- A statistical add-in program. The Handbook templates or example spreadsheets and spreadsheet macros are written using @Risk version 3.5e form Palisades. Although you may be able to alter them to use a different @Risk version or a different statistical add-in program, we are not prepared to support this activity.
- A database program is desirable but may not be necessary. Depending upon the number of words placed in the fields, there is no reason why you cannot use a spreadsheet for managing a component database.
- For running IRRAS, a DOS session with ANSI.SYS is required. Other requirements are found in the IRRAS Reference Manual, which is located on the Handbook CD-ROM.

APPENDIX

TOOLS 

The following documents are found on the Handbook CD-ROM. These documents are in .pdf (portable document format). To access these documents, follow the instructions in the \READER subdirectory to load the appropriate version of the Adobe Acrobat Reader onto your machine; then use the Reader on-line help to guide you in loading and viewing the appropriate .pdf file(s). [The latest version of the reader that was available at the time the disk was published is provided for Microsoft Windows operating systems. Readers for other formats and subsequent versions of the reader are available from Adobe at www.adobe.com.]

C.1 □ APPENDIX E

Appendix E is a listing, in .pdf format, of the NERC-GADS cause codes. Its contents are shown in this book in Appendix E. The appendix itself is located on the Handbook CD-ROM in the /C_CODES subdirectory. To use the appendix, load app_e.pdf into the Reader.

C.2 □ FAULT TREE HANDBOOK

This document was originally published by the US NRC as NUREG-0492. Although it is an excellent tutorial on fault trees, it is no longer in print. The Fault Tree Handbook is located on the Handbook CD-ROM in the /FTHB subdirectory. To use the Fault Tree Handbook, load fthb.pdf into the Reader.

C.3 □ IRRAS V. 4.0

IRRAS, the Integrated Risk and Reliability Analysis Software, was originally published by the US NRC. Version 4.0 was the last DOS-based version of IRRAS. If you find IRRAS useful, you might want to obtain SAPHIRE, an updated Windows-based version of the software, from Lockheed-Martin. You can contact them through their Web site at http://sageftp.inel.gov/saphire/saphire.asp.

IRRAS must be run in a DOS session that has ANSI.SYS loaded. You can do this using an appropriate line in the AUTOEXEC.BAT file in native DOS or with a notation in the IBM OS/2 settings notebook for the DOS window or full-screen session. Microsoft Windows 95 users will need to follow the instructions that are located in the IRRAS Reference Manual to run IRRAS. You might need to consult your Windows 95 documentation for further instructions.

The IRRAS files are located on the Handbook CD-ROM in the \IRRAS subdirectory. Follow the instructions in the IRRAS Reference Manual for loading IRRAS onto your machine. Use the IRRAS Tutorial and Reference Manual to help you learn to use IRRAS.

C.4 □ IRRAS V. 4.0 TUTORIAL

This document was originally published by the US NRC as NUREG 5813 Vol. 2. It introduces IRRAS, an integrated risk analysis program, and provides step-by-step instructions in its basic feature. To use the Tutorial, load tutorial.pdf into the Reader.

C.5 □ IRRAS V. 4.0 REFERENCE MANUAL

This document was originally published by the US NRC as NUREG 5813 Vol. 1. It exhaustively describes the IRRAS command and data structures, syntax and use. To use the Reference Manual, load reference.pdf into the Reader.

APPENDIX

HUMAN FACTORS

D.1 □ INTRODUCTION

Previous documents in the ASME Risk-Based Inspection series mention human error but do not identify specific tools for dealing with it. This appendix introduces two checklists that will help you to calculate human error probabilities for a procedure within a system. Checklist A (Section D.A) asks questions that will help you to identify which human factors apply in a procedure. Then, based on the answers to the Checklist A questions, Checklist B (Section D.B) assigns the appropriate human error probability (HEP) and also suggests other human factors that could be applied to reduce the HEP.

These two checklists are applied to four sample cases. All four cases show that human error is just another system "component" that can fail. It therefore can be inserted in any of the previous quantitative methods along with other components. However, if human error is included as a component, then its consequences must also be analyzed.

Calculating a HEP can involve more than just reading a value from Checklist B. For a given procedure, you may have to divide the procedure into sub procedures and apply the checklists to each sub procedure before you can calculate the overall HEP for the procedure.

The first two cases assess human error as one of several system components that can fail. These cases apply a procedure that is similar to fault tree analysis (see Section 8.5). The last two cases assess human error as a single-component system failure.

D.1.1 □ Constraints and Opportunities

Although Checklists A and B can help you to assess human error probabilities for many situations, they do have limitations:

Human Factors (Checklist A)

- The three types of human factors (feedback, dependency, and adjustors) that are considered in Checklist A represent action errors (subject aims at the right goal but fails to get there) more accurately than they represent planning errors (subject aims at the wrong goal and does get there). Action errors are based more on memory and attention failures whereas planning errors are based more on problem solving failures and incorrect procedures. With unintended violations, the errors are likely to be behavior-based failures.
- Feedback and dependency can correct an error as long as the error is corrected before the error leads to a hazard. Error probabilities can be adjusted lower with training, better procedures, or an improved operator interface. They can be adjusted higher by an adverse physical environment (uncomfortable clothing, temperature, or humidity; confined or complex space; poor lighting; excess noise, etc.).

Human Error Probabilities (Checklist B)

- These HEP values are also based upon action errors rather than planning errors.
- The HEP values in Checklist B are generic rather than site-specific. You could use expert elicitation (Section 9.3.1) to get site-specific HEP values.
- Over time, the HEP values in Checklist B are considered constant rather than dynamic.
- Because these HEP values were created for the nuclear power industry, they involved continuous processes, however, they considered both standby and operating components. Therefore, they also apply to industries that use batch processes.
- These HEP values represent action(s) that could lead to an adverse consequence rather than response action(s) (action(s) that respond to an emergency).
- Response HEP values approach 1.0 when there is little time to respond and are reduced when more response time is available. Response HEP values are therefore only a function of the available time and are outside the scope of this appendix.
- The HEP values represent individuals rather than teams.

D.2 □ HEP CALCULATION PROCEDURE

D.2.1 □ Preliminary Steps

For system failures that involve more than one human error and/or component failure:

1. Describe the system in terms of equipment and people.
2. Describe the consequence(s) in terms of scenario(s) and their negative outcome(s).
3. Translate the scenario(s) into a fault tree to identify procedures that are possible sources of human error.

D.2.2 □ Assess Human Error Probability

For all system failures:

1. Translate a procedure of interest into a timeline that identifies each component task.
2. Translate the component tasks into an event tree.
3. Identify the negative outcomes.
4. For each component task on the event tree, use Checklist A to identify which human factors apply:
   a) Identify the number of components in the procedure (question 3A), the type of system (series, parallel) (question 3Ba), and the dependency between/among components (zero, high, complete) (question 3Bb).
   b) Rate adjustors, such as procedures, training, or operator interfaces (adequate or inadequate) (question 2).
   c) Look for compelling feedback (question 1).
   d) Look for obvious feedback (test) (question 4).
   e) Look for subtle feedback (verify or check status with a check-off) (question 5).
5. Guided by the human factors that were identified on Checklist A, use Checklist B to assign a HEP to each task.
6. Calculate the HEP for each outcome in the procedure.
7. State your findings and recommendations.

D.3 □ HUMAN ERROR CHECKLISTS

D.3.1 □ Introduction

The nuclear power industry has developed a standard method for assessing human error that can be applied to fossil fuel-fired power plants and other industrial facilities.

In August 1983, the *Handbook of Human Reliability Analysis with Emphasis on Nuclear Power Plant Applications*, NUREG/CR-1278, [Swain, 1983] was issued. However, it is long and cumbersome. In February 1987, the *Accident Sequence Evaluation Program Human Reliability Analysis Procedure*, NUREG/CR-4772, [Swain, 1987] was issued as a shortened version of THERP. The principles and logic that underlie the "normal operation" portion of ASEP are not very complex or intricate. Unfortunately, the ASEP procedures and tables might lead you to believe otherwise. Even though deriving the human error probability (HEP) for a given procedure requires only 6 attributes, the ASEP procedure is confusing, and the tables have numerous footnotes and exceptions. This has detracted from ASEP and made HEP calculations less clear.

To enhance the usefulness of ASEP for the HEPs that are presented in this section, a simplified version was needed. The goal was to develop checklists that are accurate in terms of the intent of ASEP, but are clearer and easier to use than the ASEP manual. Checklist A and Checklist B are the result. These modified checklists will also serve as raw data collection charts and present their results, thereby documenting all the ratings and at the same time allowing "What-If" musings. Given a task, Checklist A asks human factor questions; the answers identify the HEP on Checklist B. Note that these checklists could easily be entered into a spreadsheet for automatic calculation and storage.

This methodology must be used intelligently and judiciously. While ASEP is intended to allow systems analysts to do human reliability analysis (HRA) with minimal support from human factors specialists, the ASEP validation effort concluded that some support and input from individuals with training in human behavior may still be required. [Swain, 1987] noted that "systems analysts who have had no formal training in human factors or psychology were able to apply the rules correctly in most cases." However, he later emphasized that, given the difficulties observed, he "strongly believes that any probabilistic risk assessment (PRA) that purports to consider the impact of human errors must include a specialist in HRA in the PRA team." He went on to say that those who intend to use the checklists should use these tools, then consult with human factors experts about the tasks that do not fall neatly into the ASEP framework or about which questions arise.

D.3.2 □ Checklist A: Human Factors in ASEP Normal Operation

Checklist A (Section D.A) elicits and records activity-related information that is associated with the human factors of feedback, dependency, and adjustors (procedures, training, and interface). The following paragraphs list and explain the 6 questions on Checklist A.

1. Compelling feedback (Do you know immediately that you must correct an error?)

   Checklist A asks "Is the wrong action fully corrected because overwhelming and fast feedback, such as signals from one or more annunciators, alarms, valve position indicators, or anything else, can lead the operator to correct the error after the task has been completed but before it leads to an undesired outcome?" If a compelling feedback is immediately available, then no further analysis is needed and the HEP is negligible.

   This question gives credit for built-in system safety features. In the face of a clear and compelling signal, it would take an intentional act of sabotage to not rectify the situation and prevent the undesired outcome from occurring.

2. Adequate adjustors (Are procedures, training, and people-machine interfaces adequate?)

   Checklist A notes "If overall attention to administrative controls, emergency and operating procedures, and training is very poor or if the human-machine interface is quite deficient, rate adequacy as 'no', otherwise rate as 'yes'." Swain's intent here is to increase the HEP if the overall human factors are particularly bad.

   As a rule of thumb, Swain suggests that the default rating should be

"Adequate." Only if it is readily apparent that an accident is waiting to happen should "Inadequate" be used.

3. Rate dependency among tasks (If a task fails, will another task fail because its related?)

   This part of Checklist A determines how much potential errors may be interrelated.

   Question 3A asks how many tasks are associated with a procedure.

   Question 3B has four sub-questions. The answers to the sub-questions on the left determine which, if any, further sub-questions need to be asked to determine the dependency among the tasks:

   a) Tasks are within the context of which type of system?

      A series system exhibits zero dependency because one or more failed tasks among any number of tasks will fail the system. Such a system would be represented in a fault tree by an "OR" gate.

      Figure D.1 is an example of a series system. Before starting the standby pump, if someone fails to open either valve A or valve B (or fails to open both valves A and B) in the pump suction line, the pump will fail. The tasks have zero dependency because failing to open either (or both) valves fails the pump. Therefore, any relation between the valves is unimportant.
      A parallel system needs to have two or more tasks fail before the system fails. Such a system would be represented in a fault tree by an "AND" gate.

      Figure D.2 is an example of a parallel system. Before starting the standby pump, if someone fails to open either valve A or valve B, the pump will work. To fail the pump, someone would have to fail to open both valves. In this case, the relationship between the two valves is important.
      The answers to these questions identify how much the task "open valve A" depends-on or relates to the task "open valve B."

   b) (For a parallel system) How much time is there between tasks on similar components?

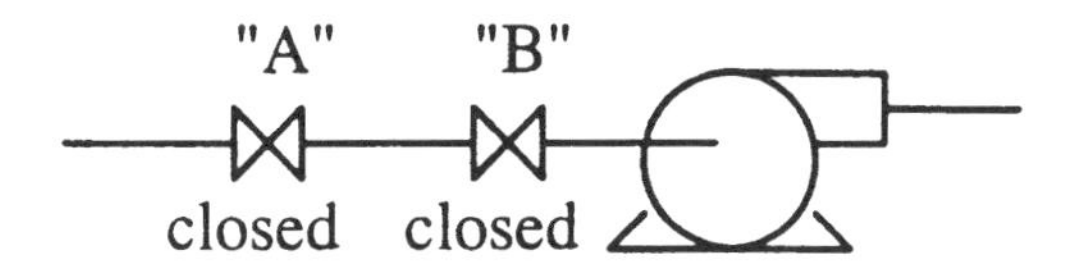

Figure D.1 Series Standby Pump System

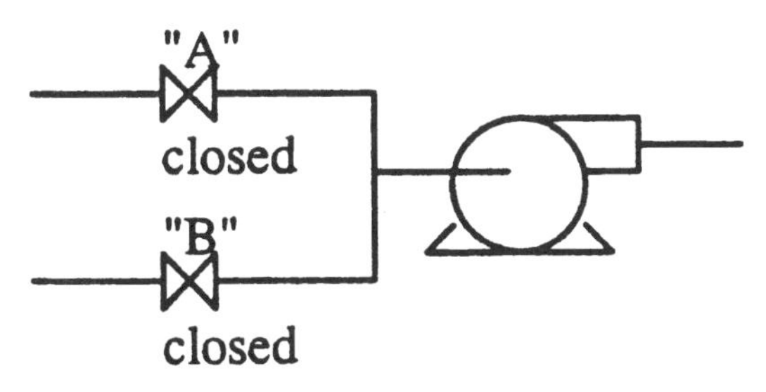

Figure D.2 Parallel Standby Pump System

If more than two minutes are available, the dependency is zero. (Whatever a person does to valve A has no relation to valve B.)

If two minutes or less are available, then sub-question 3c applies;

c) (For a parallel system) For tasks on similar components, how close together are the components?

If similar components are in different areas, a task on one component is independent of a task on the other and the dependency is zero. (If someone fails to open either valve A or valve B, they will open the other valve.)

If similar components are separated 4 feet or less, a task on one component is completely dependent on a task on the other. (If someone fails to open valve A or valve B, they will also fail to open the other valve.)

If similar components are separated more than 4 feet but are located in the same general area, then sub-question 3d applies.

d) (For a parallel system) Written sign-offs are made for each task?

If the answer is yes, then the tasks have zero dependency on each other. (If someone fails to open either valve A or valve B, they will open the other valve.)

Otherwise, the dependency is high. (If someone fails to open either valve A or valve B, they may fail to open the other valve.)

4. Obvious feedback (Was the system tested Post-Maintenance or Post-Calibration?)

   This section asks "Can maintenance or calibration error(s) be identified and corrected by correctly testing the system before returning it to service?"

5. Subtle feedback (Is a checklist used when verifying or checking (a) task(s) result(s)?)

   This section looks for a checklist.

5A. Is the task(s) result immediately verified with a written checklist?

Up to 3 sub-questions are asked:

a) Second individual personally verifies (a) task(s) result(s)?

   If the answer is yes, then sub-question 5c applies.

   If the answer is no, then sub-question 5b applies.

b) Performer checks task(s) at different time and place?

   If yes, then sub-question 5c applies.

   If the answer is no, the final result is No. Note that this makes sense in control room applications where manual actions in the plant may result in on/off, open/shut, or indicator value changes in the control room.

c) Is a written check-off used?

   If the answer is yes, then the result is considered verified;

   If the answer is no, then the result is not considered verified.

5B. Is the status of a task(s) result checked regularly with a written checklist?

This section consists of two sub-questions:

a) Is the status checked regularly per batch or shift?

If the answer is yes, then sub-question b applies;

If the answer is no, then the final rating is No. In a control room setting, a regular walk-through of the facility is an important way of catching problems before they become serious. Strange noises, leaks, vibrations, lack of indication, or any other abnormal condition that can only be observed directly can indicate serious problems that can be mitigated if they are detected early.

b) Is a written check-off used?

If yes, then the final rating is Yes;

If no, then the final rating is No.

D.3.3 □ Checklist B: Human Error Probability

Checklist B allows for looking up the appropriate HEP, given the answers that are marked on Checklist A. The Checklist A ratings are used to select a mean HEP value with an error value. The error value defines the HEP range by dividing and multiplying the mean value. For example, 0.05 (5) has a mean HEP of 0.05 and a HEP range of 0.01 to 0.25.

VALUE OF CHECKLIST B

Having all the values available and the attribute combinations that lead to the values makes it easy to do "what-if" comparisons and informal sensitivity studies. For example, one can easily see the impact that status checks, verification, and Post-Maintenance/Post-Calibration tests have on the HEPs.

Another advantage of the values shown in Checklist B is that they are all mean values rather than the median values that are reported in the ASEP document. This facilitates use of these values in PRAs and in other risk analysis techniques.

ATTRIBUTES

This section describes how the human factors of dependency among tasks, feedback, and adjustors alter basic human error probabilities (BHEPs) and generate a HEP for a procedure.

A BHEP is 0.02 for forgetting to act, 0.01 for doing an act incorrectly, or 0.03 for combining both types of errors.

Dependency is described first because event trees and equations explain how to calculate a HEP for a task from a BHEP and how to apply the differences between various dependencies.

The following additional variations to the BHEPs are shown in the order of their ability to reduce the error.

Compelling feedback reduces a BHEP by about 0.001.

Obvious feedback from a test reduces a BHEP by about 0.01.

Subtle feedback from verification and status checks reduces a BHEP by about 0.1.

Adjustors such as training, procedures, and interface reduce a BHEP by only about 0.5.

The highest HEPs are located in the rows that are defined by repeated no answers to compelling feedback, adjustors, test, verification, and status check. In these cases, the HEP approximates the BHEP.

Dependency and the Number of System Components

The three major rows within each of the two "Adjustor" rows represent the three levels of dependency that are considered. There may be up to 5 sub-columns that represent the different HEP given the number of tasks in either the series or parallel systems.

As the number of zero dependency tasks increases (e.g. one failed task can fail the system), the HEP increases (the chance increases of neglecting one of the in-series tasks).

For a series system with n independent tasks, HEP $\approx$ BHEP $\times$ n. For Figure D.1 in Section D.3.2, with BHEP = 0.03 and n = 2, BHEP $\times$ n approximates the following event tree (figure D.3a):

For a parallel system with n independent tasks, HEP $\approx$ $(\text{BHEP})^n$. For Figure D.2 in Section D.3.2, with BHEP = 0.03 and n = 2, $(\text{BHEP})^n$ approximates the following event tree (figure D.3b):

In the high dependency row (parallel systems in the same general area but not within 4 feet, but which must be handled within 2 minutes of each other and for which there are not written sign-offs for each component), the HEPs decrease with the number of tasks.

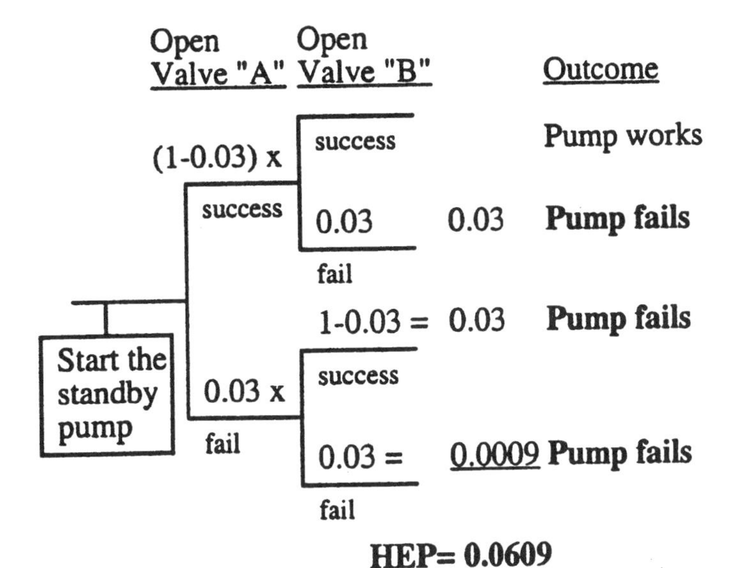

Figure D.3a Event Tree for a Series System, Independent Tasks

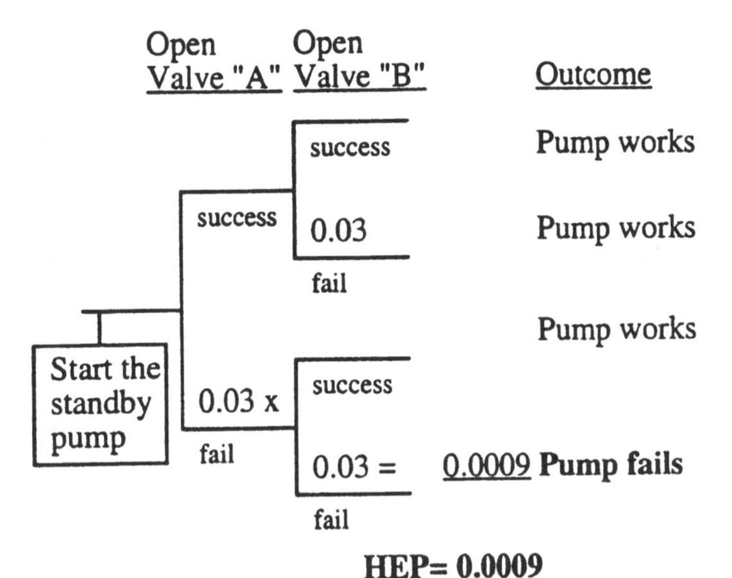

Figure D.3b Event Tree for a Parallel System, Independent Tasks

For a parallel system with n highly dependent tasks, HEP ≈ BHEP × $(0.5)^{n-1}$. For Figure D.2 in Section D.3.2, with BHEP = 0.02 and n = 2, BHEP × $(0.5)^{n-1}$ approximates the following event tree (figure D.3c):

For systems with complete dependency, there is only one HEP value regardless of the number of tasks involved.

For a parallel system with n completely dependent tasks, HEP ≈ BHEP ×

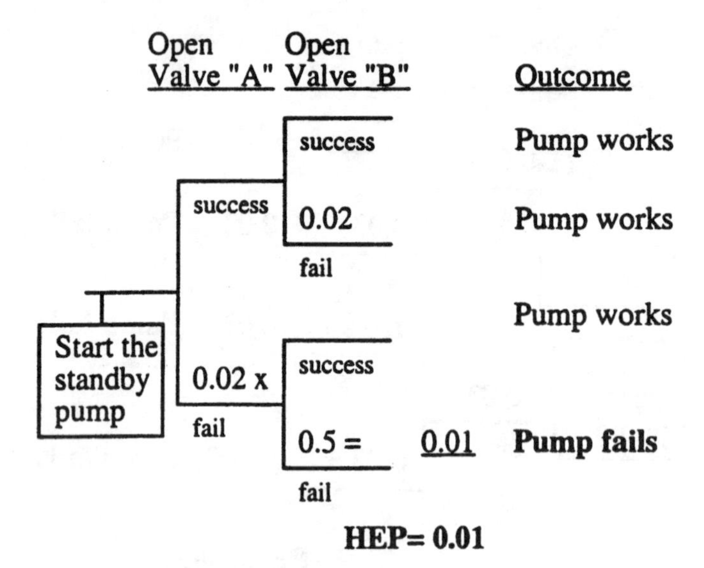

Figure D.3c Event Tree for a Parallel System, Highly Dependent Tasks

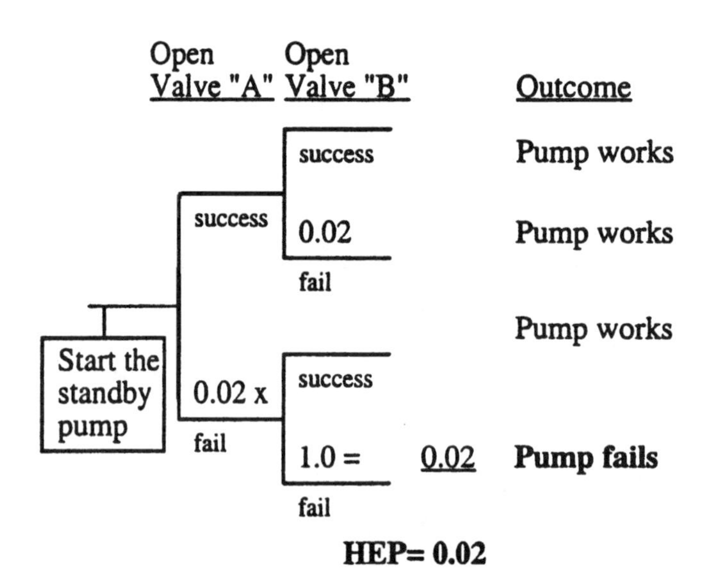

Figure D.3d Event Tree for a Parallel System, Completely Dependent Tasks

$(1.0)^{n-1}$. For Figure D.2 in Section D.3.2, with BHEP = 0.02 and n = 2, BHEP × $(1.0)^{n-1}$ approximates the following event tree (figure D.3d):

Compelling Feedback

If there is a compelling signal, the HEP is negligible (i.e., less than 1.0 E-5). No need to go any further.

For a series system with n independent components and a compelling

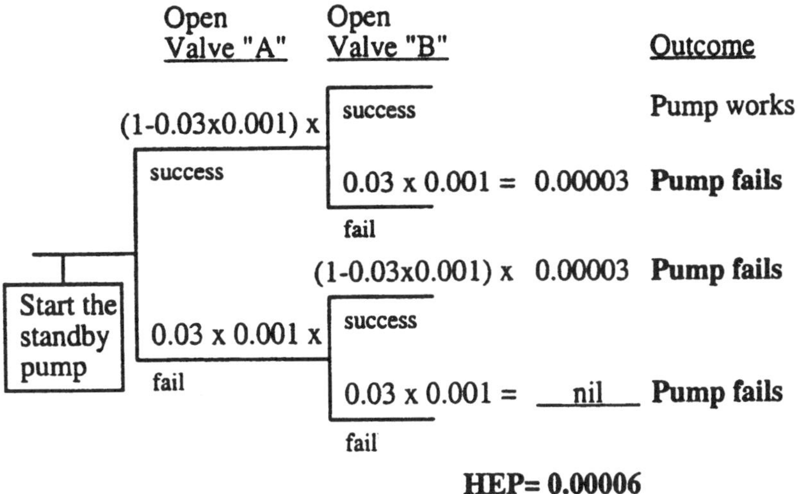

Figure D.3e Event Tree for a Series System, Independent Tasks, with Compelling Feedback

signal such as a positioner switch on valves, HEP $\approx$ (BHEP $\times$ 0.001) $\times$ n. For Figure D.1 in Section D.3.2, with BHEP = 0.03 $\times$ 0.001 and n = 2, (BHEP $\times$ 0.001) $\times$ n approximates the following event tree (figure D.3e):

For a parallel system with n independent components, HEP $\approx$ (BHEP $\times$ 0.001)^n.

For a parallel system with n highly dependent components,

$$\text{HEP} \approx (\text{BHEP} \times 0.001) \times (0.5)^{n-1}$$

For a parallel system with n completely dependent components,

$$\text{HEP} \approx (\text{BHEP} \times 0.001) \times (1.0)^{n-1} = \text{BHEP} \times 0.001$$

Obvious Feedback (Test the system Post-Maintenance or Post-Calibration)

Each Dependency row is sub-divided by "PM/PC tests" into a yes and a no row. The tests correct the previous task and reduce the BHEP by about 0.01.

For a series system with n independent components and a test, HEP = (BHEP $\times$ 0.01) $\times$ n. For Figure D.1 in Section D.3.2, with BHEP = 0.03 $\times$ 0.01

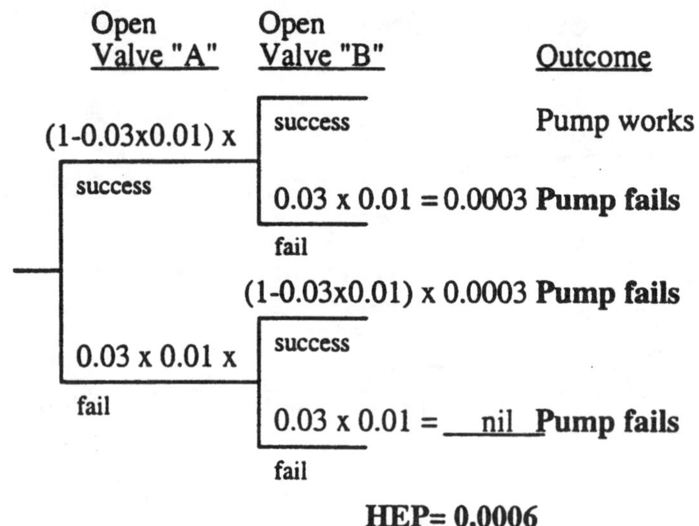

Figure D.3f Event Tree for a Series System, Independent Tasks, with a Test

and n = 2, (BHEP × 0.01) × n approximates the following event tree (figure D.3f):

For a parallel system with n independent components and a test,

$$HEP \approx (BHEP \times 0.01)^n$$

For a parallel system with n highly dependent components and a test,

$$HEP \approx (BHEP \times 0.01) \times (0.5)^{n-1}$$

For a parallel system with n completely dependent components and a test,

$$HEP \approx (BHEP \times 0.01) \times (1.0)^{n-1} = BHEP \times 0.01$$

Subtle Feedback (Written checklist is used to verify task results or to check system status)

Each of the PM/PC Test rows are further subdivided by "Verified" into "Yes" and "No" rows. Similarly, the "Verified" rows are subdivided by "Status Checks" into "Yes" and "No" rows. "Verify" and "Status checks" correct the previous task and reduce the BHEP by about 0.1. However, if there is a PC/PM test, "Verify" and "Status Check" do not further reduce BHEP.

For a series system with n independent components and verification or checks,

$$HEP \approx (BHEP \times 0.1) \times n$$

For a parallel system with n independent components and verification or checks,

$$HEP \approx (BHEP \times 0.1)^{n}$$

For a parallel system with n highly dependent components and verification or checks,

$$HEP \approx (BHEP \times 0.1) \times (0.5)^{n-1}$$

For a parallel system with n completely dependent components and verification or checks,

$$HEP \approx (BHEP \times 0.1) \times (1.0)^{n-1} = BHEP \times 0.1$$

Adjustors (Procedures, training, people-machine interfaces)

The 2 major rows represent values for adequate and inadequate adjustors. In all cases, the corresponding values are slightly higher for the inadequate adjustors.

For a series system with n independent components and human factors,

$$HEP \approx (BHEP \times 0.5) \times n$$

For a parallel system with n independent components and human factors,

$$HEP \approx (BHEP \times 0.5)^{n}$$

For a parallel system with n highly dependent components and human factors,

$$HEP \approx (BHEP \times 0.5) \times (0.5)^{n-1}$$

For a parallel system with n completely dependent components and human factors,

$$HEP \approx (BHEP \times 0.5) \times (1.0)^{n-1} = BHEP \times 0.5$$

D.4 □ EXAMPLES

This section illustrates the procedure by calculating the HEP within two multi-component systems and as a single "component" in two types of inspections.

- The first multi-component system example involves a fuel hose that creates a fire if it fails.
- The second multi-component system involves water leaking into a system during shutdown and then rupturing piping during startup.
- The first single-component example involves a visual maintenance inspection. If the visual inspection fails, a steam drum's efficiency will decrease.
- The last example involves post-repair inspections that are conducted to verify weld integrity. If these inspections fail, a repaired component could leak or rupture after unit restart. A unit shutdown to repair the leak would be required. A rupture could have serious safety consequences. (Note that this example, like the others, addresses and illustrates only the human factors portion of the failure probability. This example does not address the probability of detection (POD) issues that you would also need to consider when you analyze a nondestructive examination issue.)

D.4.1 □ Human Error(s) (Constant Rate) in a Multi-Component System

This section considers two of the systems in the typical fossil fuel-fired power plant that is shown in Figure D.4a. One case involves the fuel oil lighter system and the other involves the reheat attemperator water supply line. Each system has more than one component and human error is one or more of these components.

CASE 1 (Fuel Oil Lighter system)

The assessment is divided into Preliminary (D.2.1) and Human Error Assessment (D.2.2) Steps.

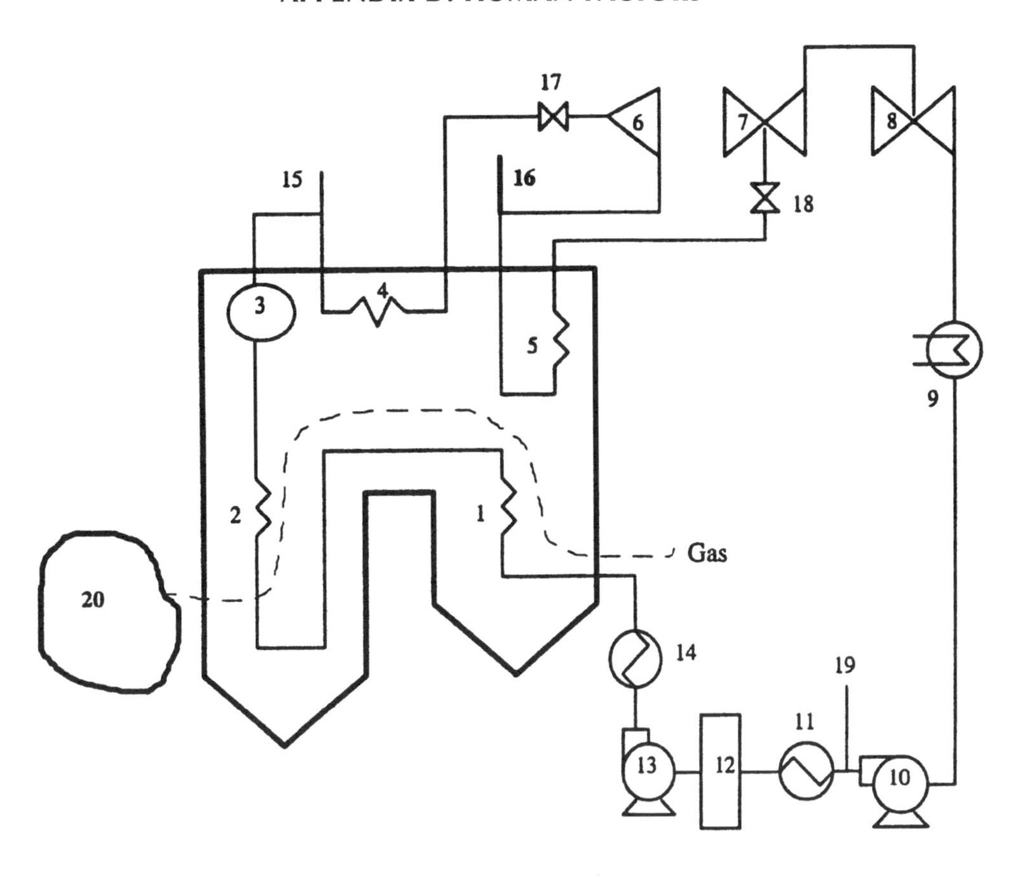

1. Economizer
2. Waterwall tubes
3. Steam drum
4. Superheater
5. Reheater
6. High pressure turbine
7. Intermediate pressure turbine
8. Low pressure turbine
9. Condenser
10. Condensate pump
11. Low pressure feed heaters
12. Deaerator
13. Boiler feed pump
14. High pressure feed heater
15. Superheater attemperator
16. Reheater attemperator
17. Stop valve
18. Intercept valve
19. Condensate makeup
20. Fuel oil lighter system

Figure D.4a Fossil Fuel-fired Power Plant Process Flow Diagram (Simplified)

Preliminary Steps

These steps help you to find where human error will contribute to system failure.

Step 1: Describe the system

Before coal can begin feeding the furnace (the area in figure D.4a that contains components 1–5), the combustor (one of 14) has to be

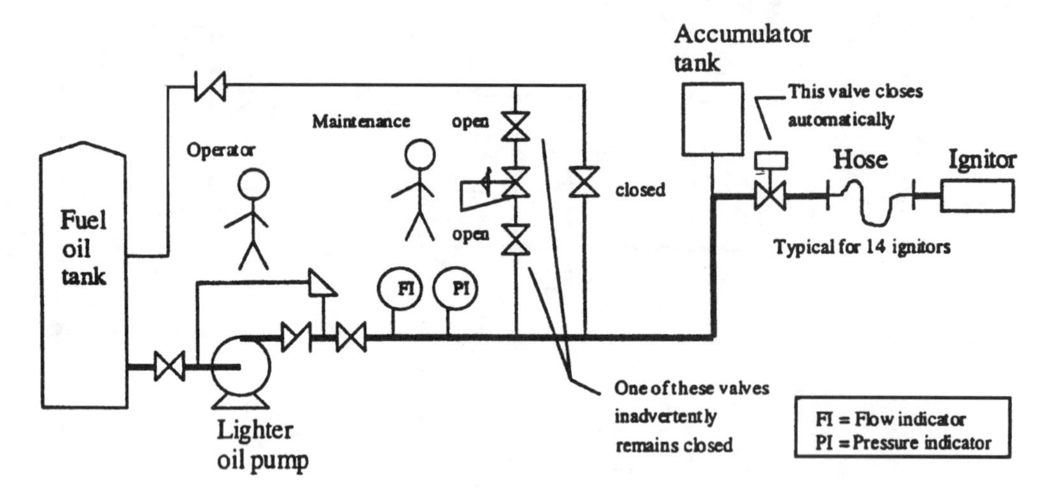

Figure D.4b Fuel Oil Lighter System Process Flow Diagram

heated to 2000°F. A combustor is heated by continuously feeding it a mixture of oil, coal, and air. The fuel oil lighter system supplies oil to heat the coal and air mixture up to 2000°F and then stops. See Figure D.4b.

A pump draws oil from a fuel oil tank and transfers it through a hose into ignitors that are located in the combustors. The oil enters that combustor as a burning spray. A flexible hose in the feed line allows the combustor to expand as it heats without stressing the piping. A valve (located in figure D.4b between the valves labeled "open") in a recirculating line regulates the pump discharge pressure by recirculating oil to the fuel oil tank. This feature is necessary because oil may flow to one or more combustors at any time during the startup.

Periodically, to fix leaks and to inspect the pressure regulator in the recirculating line, maintenance personnel close the plug valves (the valves labeled "open" in Figure D.4b) in the fuel oil line. The plug valves are closed by inserting a portable handle into a notch and then turning the plug. A valve's position is difficult to see.

Step 2: Describe the consequence(s)

The consequence(s) are described by describing a scenario that has adverse outcomes.

Scenario

1. Either a plug valve inadvertently remains closed after a maintenance inspection or the regulator fails closed.
2. An operator starts the fuel oil pump to heat a combustor.
3. Excess pump pressure, caused by the lack of recirculation flow, ruptures a fuel oil hose.
4. Fuel oil sprays onto a burner and ignites outside the furnace.
5. The fire protection system fails.

Consequence and Severity

Lost production and property damage when oil creates a fire in the burner area. (One incident required 3 to 4 weeks of working 24 hr/day to get back on line). Possible safety-related consequences.

(Step 3): Translate the scenario into a fault tree to find human error opportunities

a. Given the above scenario, develop the fault tree and stop when the basic events are all human errors or components. See Figure D.5. For more information about fault trees, see The Fault Tree Handbook (Appendix C.1) For automated fault tree analysis, see IRRAS, its manual and tutorial (Appendices C.2–C.4).

   For this example, the human errors are identified as two active errors and three latent errors. Active errors lead immediately to the adverse outcome(s) whereas the latent errors lead to adverse outcome(s) but not immediately.

b. For each basic event that is a human error:
   i. Identify the component task(s) in a procedure
   ii. Use Checklist A to identify the human factors that are involved in the task
   iii. Use Checklist B to assign a human error probability.

c. Given the basic event probabilities, calculate the fault tree.

Figure D.5 shows the results of the above 3 steps. The remainder of this section describes how the human error basic event probabilities in Figure D.5 are calculated.

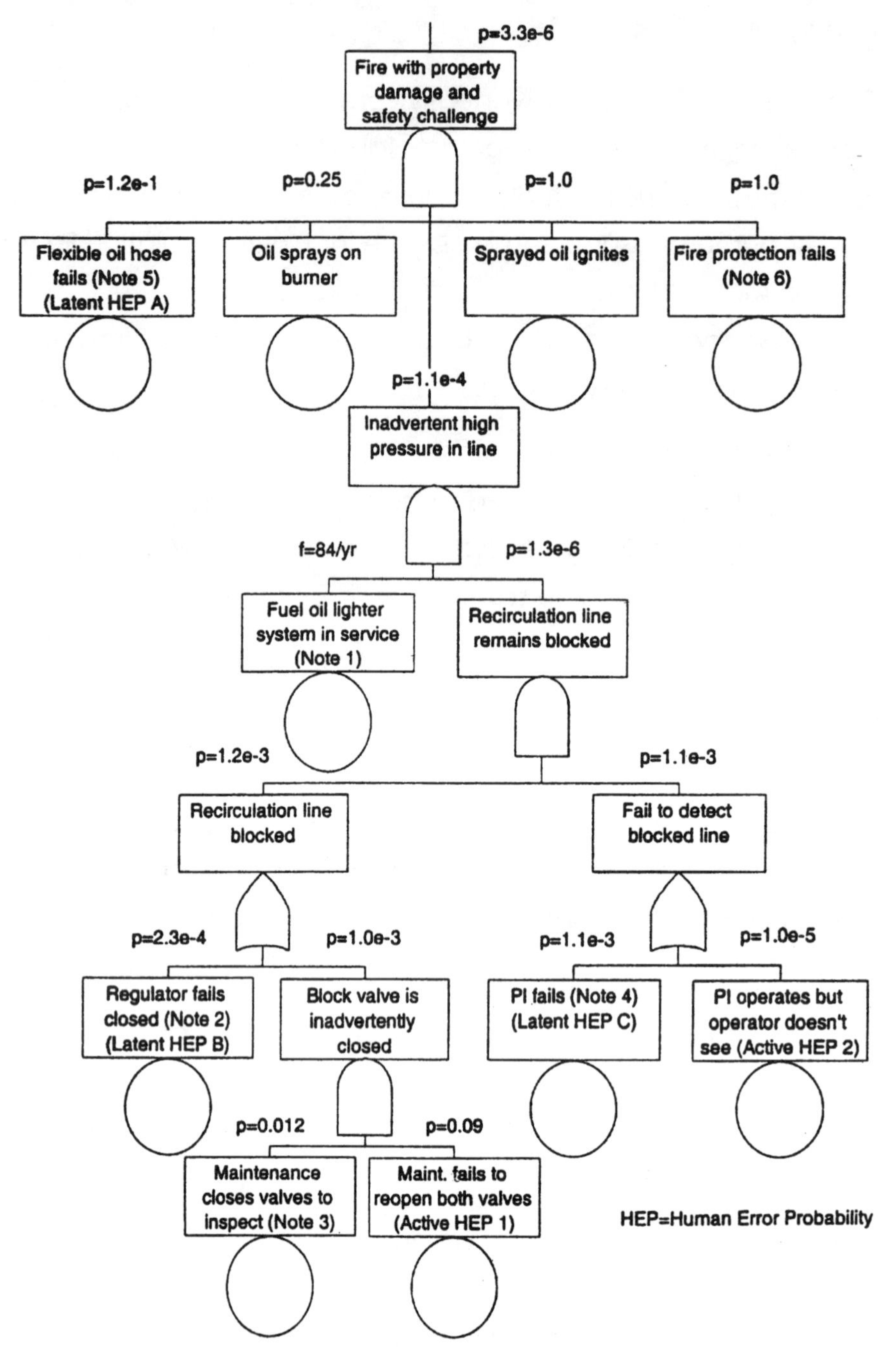

NOTES:

1. f= 84/yr = 6 starts/yr-cyclone(burner) x 14 cyclones
2. 2. 2.3 E-4 = 0.019/yr (yr/84) This is tested every time an ignitor starts.
3. 0.012 = inspect 1/yr x yr/84 startups
4. 4. 1.1 E-3 = 0.090/yr (yr/84) This is tested every time an ignitor starts.
5. 1.2 E-1 = 0.7/yr x (yr/6) This is tested every time a cyclone starts.
6. No fire protection available

Figure D.5 Fuel Oil Lighter System Fault Tree

Human Error Assessment Steps

This scenario contains five procedures that are human error opportunities. Two of the human errors are active. Active human errors contribute to an immediate consequence. The other three are latent. Latent human errors contribute to a delayed consequence; the error is undetected for a long time (e.g. hours, days, months, or years) until one or more other components fail.

1. Maintenance fails to open block valve (Active HEP 1)
2. PI operates but operator fails to detect high pressure (Active HEP 2)

3a. Flexible hose fails (Latent HEP A)
3b. Regulator fails (Latent HEP B)
3c. PI fails (Latent HEP C)

This section assesses Active HEP 1 and Active HEP 2 but not Latent HEP A, Latent HEP B, and Latent HEP C. These "latent" values represent equipment failure rate with no errors during their inspection procedures. To estimate these values, the equipment inspection procedures would have to be assessed.

Active HEP 1

(Step 1) Translate the written procedure to a timeline

The procedure for Active HEP 1 has two tasks.

Task 1 = Close 2 valves.

Task 2 = Reopen both valves.

(Step 2) Translate the tasks into an event tree

The tasks in Step 1 are shown as an event tree in figure D.6a. Because this procedure has two tasks, the event tree has a dedicated line for success and failure for each task.

(Step 3) Identify negative outcome(s)

For Task #1, if an operator fails to close both valves, then if the fuel pump starts, an operator could be sprayed. For Task #2, if an operator fails to reopen both valves, then the pump head could create a high pressure. Note that the outcomes differ. In this case, high pressure is the outcome of interest.

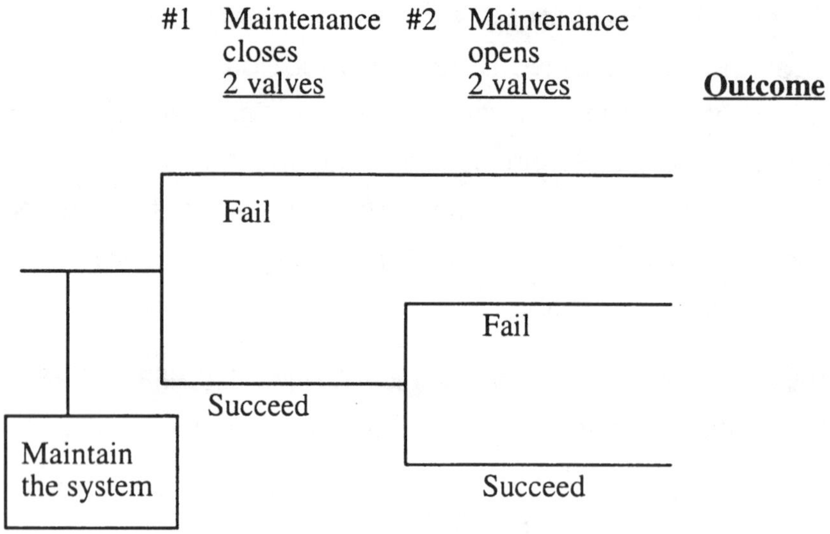

#1
Maintenance closes 2 valves
#2
Maintenance opens 2 valves
Outcome
Fail
Fail
Succeed
Succeed
Maintain the system

Figure D.6a Fuel Oil Lighter System, Active HEP 1, Step 2

#1
Maintenance closes 2 valves
#2
Maintenance opens 2 valves
Negative Outcome
Potential spill
Fail
High pressure
Fail
Succeed
Succeed
Maintain the system

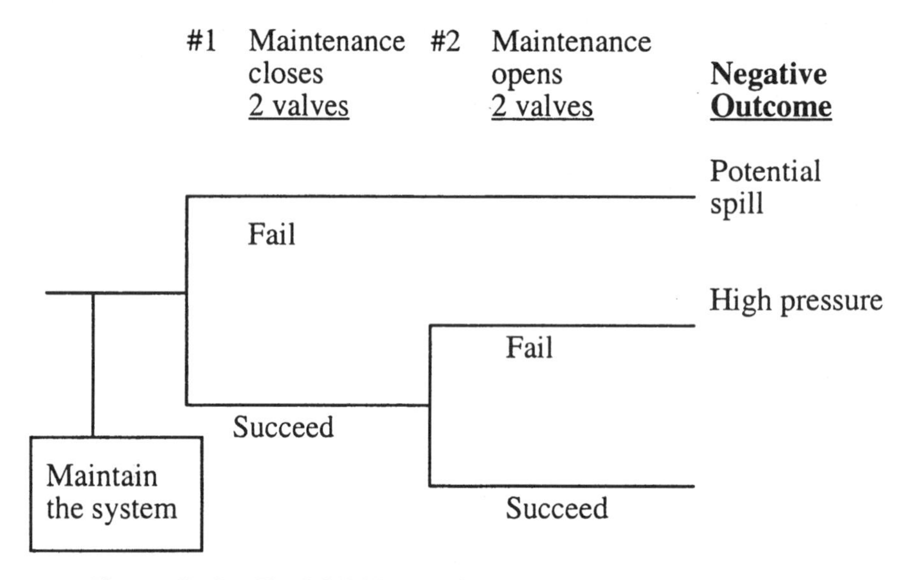

Figure D.6b Fuel Oil Lighter System, Active HEP 1, Step 3

(Step 4) For each task on the Event Tree, use Checklist A

For Task #1 and #2, go to Checklist A and answer the questions. The expected answers are listed beneath Task #1 and Task #2 in Table D.1.

(Step 5) From Checklist "A" answers, use Checklist "B" to assign HEPs See **Figure D.6c**.

(Step 6) Calculate the HEP for a particular outcome in the procedure

TABLE D.1 FUEL OIL LIGHTER SYSTEM, ACTIVE HEP-1, STEP 4

Question	Task #1	Task #2
3A	2	2
(reason)	*one task for each valve*	*one task for each valve*
3Ba	series	series
(reason)	*oil spills if either valve remains open*	*high head if either valve stays shut*
3B	Zero	Zero
2	Yes	Yes
1	No	No
(reason)	*cannot see valve position*	*cannot see valve position*
4	No	No
5Aa	No	No
5Ab	No	No
5A	No	No
5Ba	No	No
5B	No	No

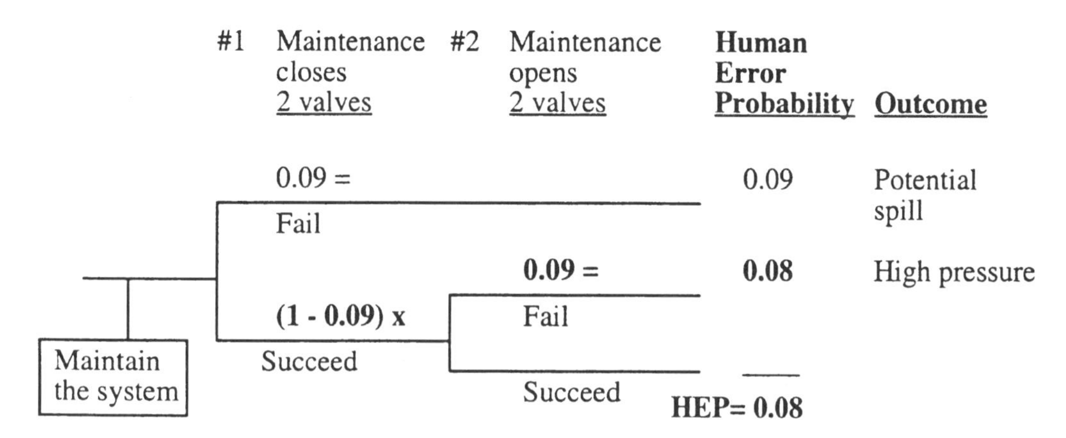

Figure D.6c Fuel Oil Lighter System, Active HEP-1, Steps 5 and 6

To calculate the human error probability (HEP) for the procedure, multiply the human error and success probabilities for Task #1 times the human error and success probabilities for Task #2 and then add the human error probabilities for each negative outcome.

In this example, to get probabilities for the two negative outcomes:

1. Copy Task #1 failure to get 0.09
2. Multiply Task #1 success (1-0.09) by Task #2 failure (0.09) to get 0.08

(Step 7) State findings and recommendations

The oil spill hazard was previously identified. Another potential hazard is exposing maintenance personnel to oil during an inspection if either the check valve in line to the tank fails or if the pump is started.

Regulator failure is a less likely maintenance error than failure to properly operate the plug valves. However, if the ignitor system is used fewer than 84 times a year, regulator failure might become more likely than valve operating error.

Using Checklist B, here are some ways to reduce the Task #2 HEP from 0.09:

- Verifying with a checklist that the valves are open will reduce the HEP to 0.01.
- A test that verifies oil can recirculate back to the tank will reduce the HEP to 0.001. (Adding a checklist to verify the status immediately after a test does NOT further decrease the HEP.)
- Installing valve positioners will reduce the HEP to 0.00001.

Active HEP 2

(Step 1) Translate the written procedure to a timeline

The procedure for Active HEP 2 has one task.

Task 1 = Detect high pressure.

(Step 2) Translate the tasks into an event tree

The task in Step 1 is shown as an event tree in Figure D.7a. Because this procedure has only one task, the event tree has a dedicated line for success and failure.

(Step 3) Identify negative outcome(s)

For Task #1, if an operator fails to detect high pressure, then the pump could create a high pressure.

(Step 4) For each task on the event tree, use Checklist "A"

The expected answers are listed in Table D.2.

(Step 5) From Checklist "A" answers, use Checklist "B" to assign HEPs

See Figure D.7b.

(Step 6) Calculate the HEP for a particular outcome in the procedure

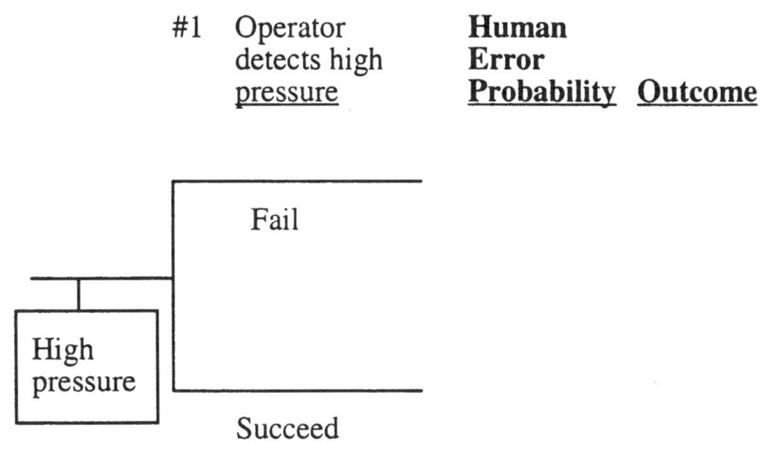

Figure D.7a Fuel Oil Lighter System, Active HEP-2, Step 2

TABLE D.2 FUEL OIL LIGHTER SYSTEM, ACTIVE HEP-2, STEP 4

Question	Answer (Task #1)	Reason
3A	1	*only one task*
3Ba	Series	*high pressure if operator fails to detect*
3B	Zero	
2	Yes	
1	Yes	*operator can see pressure gauge*

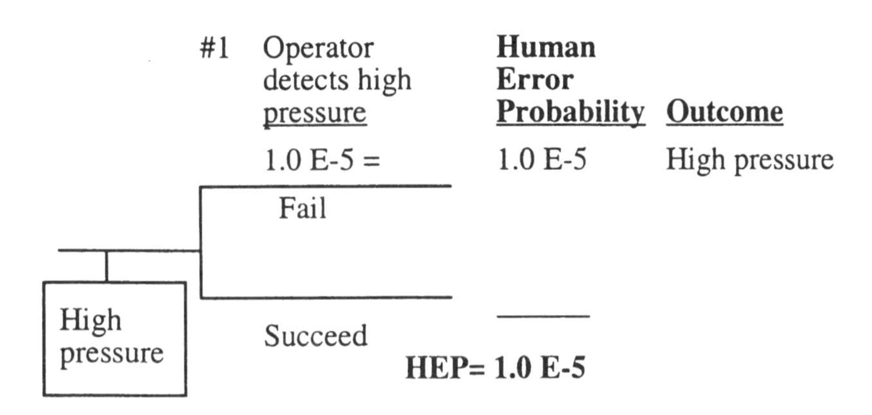

Figure D.7b Fuel Oil Lighter System, Action HEP-2, Steps 3 to 6

To calculate the human error probability (HEP) for this procedure, Task #1 is the procedure with one negative outcome.

In this example, to get the probability for the one negative outcome:

1. Copy the Task #1 Human Error Probability to get 1.0 E-5.

(Step 7) State findings and recommendations

The pressure indicator is more likely to fail than an operator is likely to not detect the high pressure.

The human error probability of 1.0 E-5 is unlikely to be further reduced.

CASE 2 (Reheat Attemperator Water Supply)

This section assesses the human errors contributing to a risk with the Cold Reheat Line. The assessment is divided into Preliminary and Human Error Assessment Steps.

Preliminary Steps

Preliminary steps help you to find the opportunities where human error will contribute to system failure.

(Step 1) Describe the system

Steam temperature is controlled by injecting water into the steam through the two attemperators (15 and 16 in Figure D.4a). If the steam temperature is too hot, creep could fail a metal component; and if the temperature fluctuates, low cycle stress fatigue could fail a metal component. To control the steam temperature from the superheater (4 in Figure D.4a), water is injected through the superheater attemperator (15 in Figure D.4a) upstream of the superheater. To control the steam temperature from the reheater (5 in Figure D.4a), water is injected through the reheater attemperator (16 in Figure D.4a) upstream of the reheater (5).

When the unit is in service, the water line valves are open and the drain valves are closed (see Figure D.9a). When the unit is shutdown, the water line is shutoff by closing two valves in the main line, and opening two drain valves (see Figure D.9b). Therefore, if a valve in the main line leaks, water will drain.

(Step 2) Describe the consequence(s)

Scenario

While shutdown, the main line valve(s) leak, an operator fails to open the drain valves. Water flows through the attemperator piping and accumulates at a low point in the cold reheat piping or reheater. When the

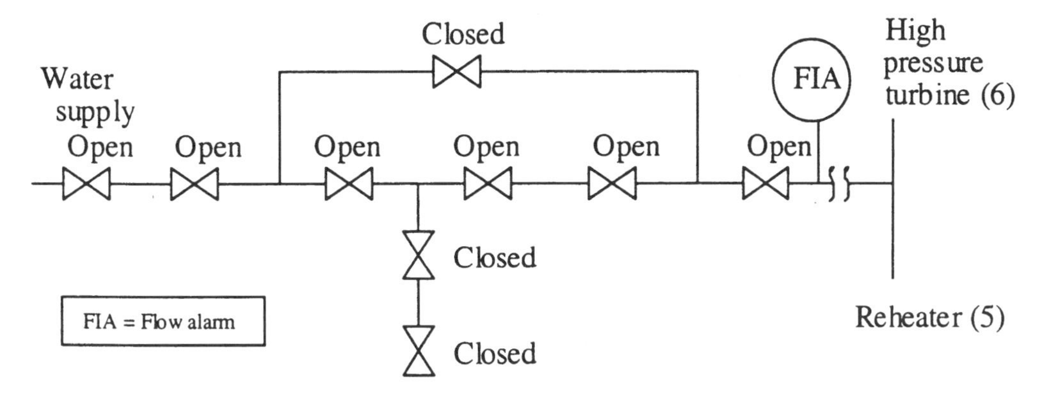

Figure D.9a Reheat Attemperator Water Supply (desired in-service lineup)

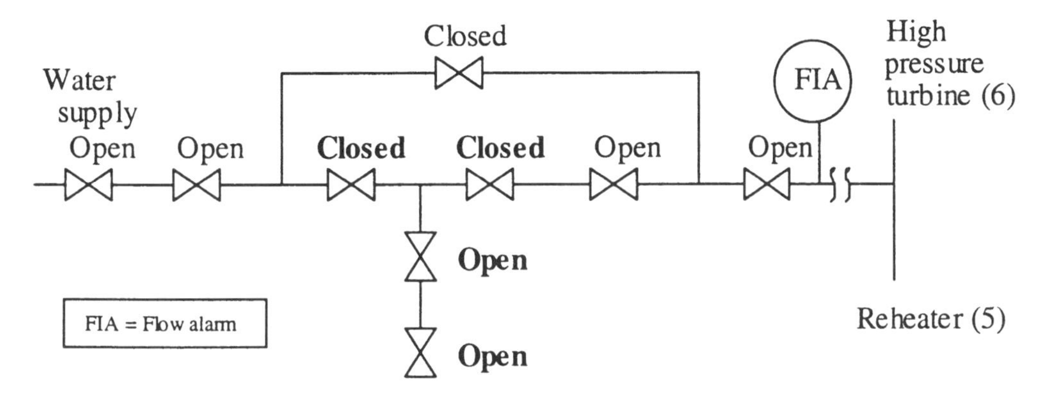

Figure D.9b Reheat Attemperator Water Supply (Desired Shutdown Lineup)

unit restarts and the turbine main stop and intercept valves open, steam propels the water along straight portions of 36-inch diameter pipe until it hits a change in piping direction, creating a water hammer, and/or enters the turbine, creating a water induction incident.

Consequences and Severity

Unit shut down for repair of water hammer damage to building steel, piping supports, and piping and/or water induction damage to the turbine.

(Step 3) Translate the scenario into a fault tree to find human error opportunities

a. Given the above scenario leading to the consequences, develop the fault tree and stop when the basic events are either human error(s) or component(s). See Figure D.10. For this example, three active errors and two latent errors have been identified.

b. For each human error basic event, identify the tasks in a procedure, use Checklist A to identify the human factors involved in the task, and then use Checklist B to assign a human error probability. This step is fully developed in the next section.
c. Given the basic event probabilities, calculate the fault tree.

Figure D.10 shows the results of the above 3 steps. The remainder of this section describes how the human error basic event probabilities in Figure D.10 are calculated.

Human Error Assessment Steps

This example contains five procedures that are human error opportunities. Three of the human errors are active. They contribute to an immediate consequence. The other 2 are latent. They contribute to a delayed consequence; the error is undetected for a long time (e.g. hours, days, months, or years) until one or more other components fail.

1. Operator closes 1 or 2 water supply valves and fails to open both drain valves (Active HEP 1)
2. Operator fails to detect and stop flow (Active HEP 2)
3. Operator fails to drain low spot(s) (Active HEP 3)
4. FIA fails (Latent HEP A)
5. Design engineer fails to install drains at low spot(s) (Latent HEP B)

This section will assess only Procedures 1 and 2 for human error. Procedure 3 is very similar to procedure 1. The tasks for Procedures 4 and 5 are unidentified.

Active HEP 1

(Step 1) Translate the written procedure to a timeline

The procedure for "Active HEP 1" has three tasks.

Task 1—to close main upstream valve.

Task 2—to close main downstream valve.

Task 3—to open 2 drain valves.

(Step 2) Translate the tasks into an Event Tree

The tasks in Step 1 are shown as an event tree in Figure D.11a. Because

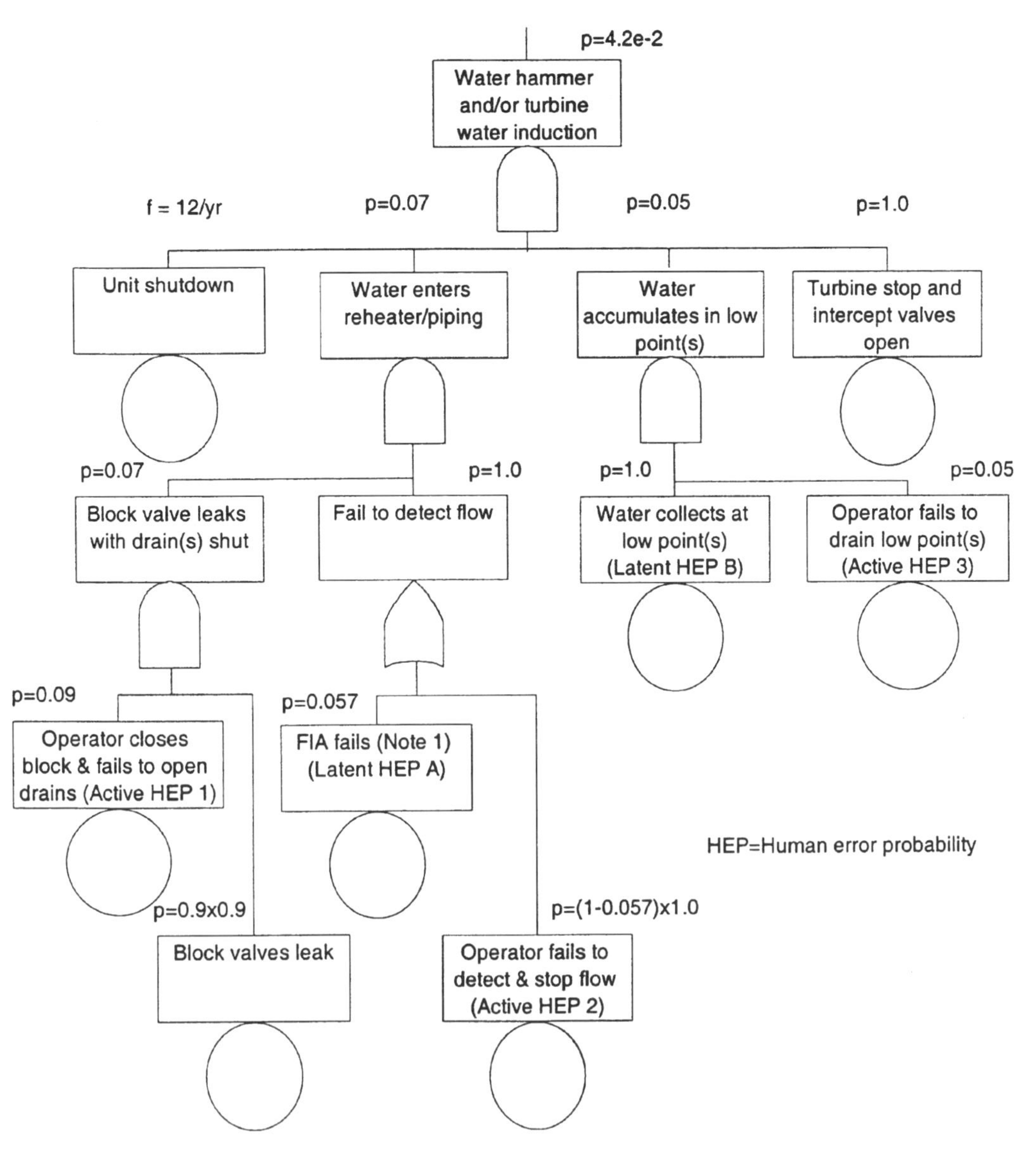

NOTES:

1. 0.057= 0.68/yr x (yr/12) however, FIA has been a nuisance for a long time; it alarms when there is no flow.

2. HEP 2 ≈ 1.0 rather than 1.0 E-5 because FIA is a nuisance. Therefore, the operators ignore it.

Figure D.10 Reheat Attemperator Water Supply Fault Tree

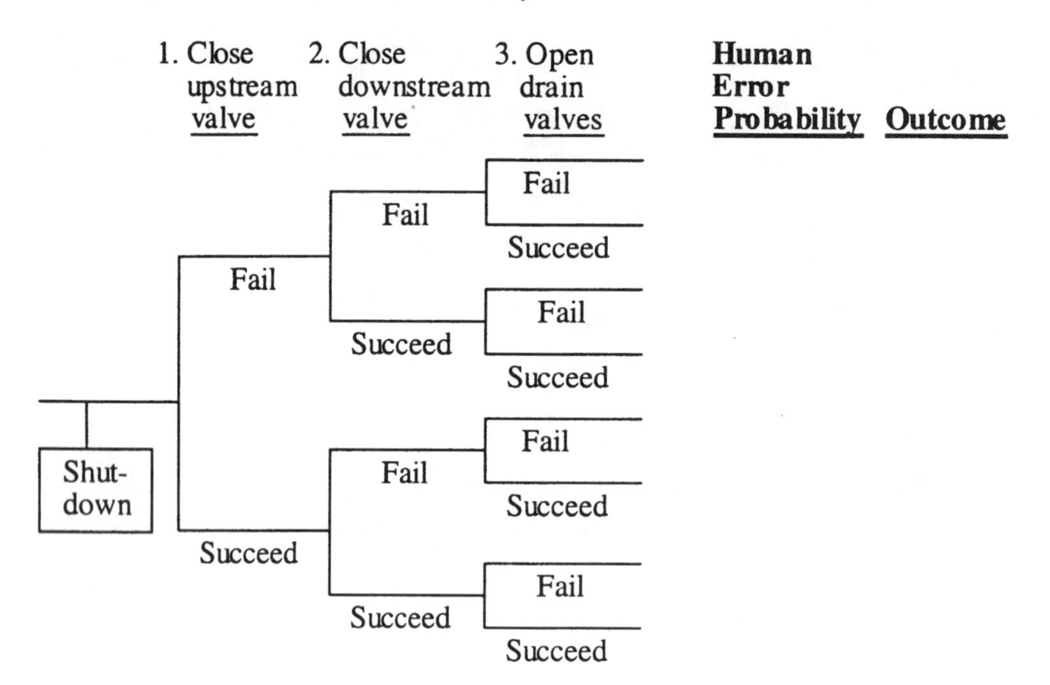

Figure D.11a Reheat Attemperator Water Supply, Active HEP 1, Step 2

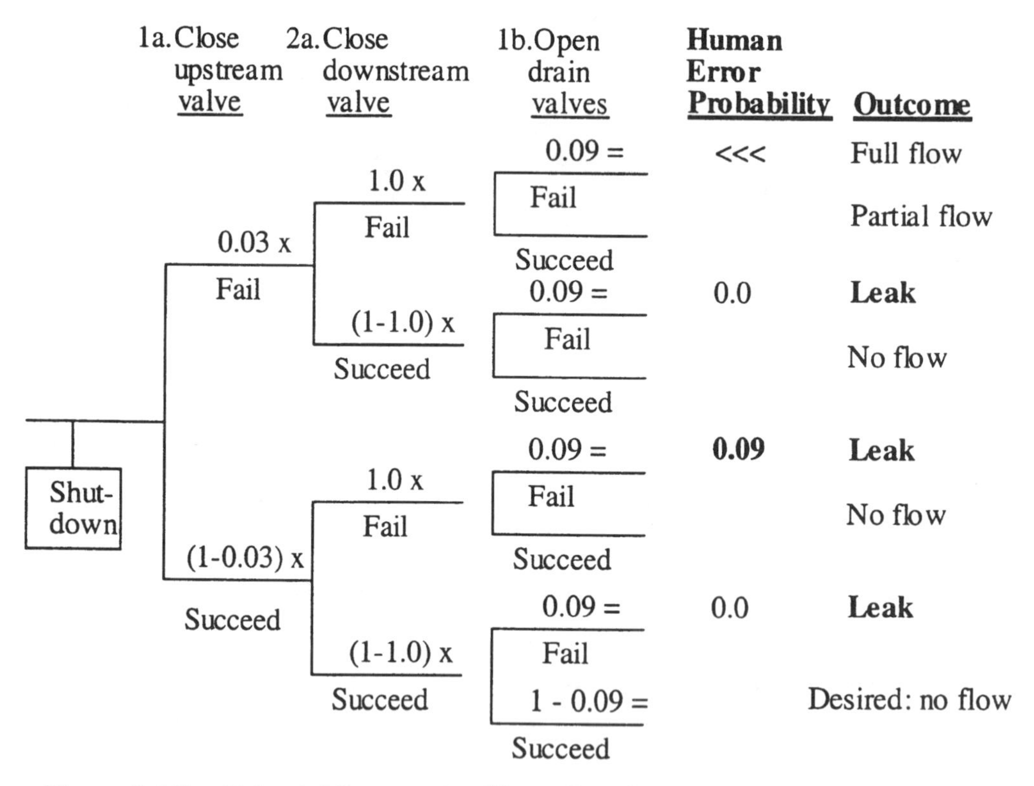

Figure D.11b Reheat Attemperator Water Supply, Active HEP 1, Steps 3 to 6

TABLE D.4 REHEAT ATTEMPERATOR WATER SUPPLY, ACTIVE HEP 1, STEP 4

Question	Task #a	Task #b
3A	2	2
(reason)	*one task for each block valve*	*one task for each drain valves*
3Ba	parallel	series
(reason)	*full flow only if both block valves remain open*	*potential water leak if either drain valve remains closed*
3b	<2 minutes	–
3c	<4 feet	–
3B	Complete	Zero
2	Yes	Yes
1	No	No
(reason)	*cannot see valve position*	*cannot see valve position*
4	No	No
5Aa	No	No
5Ab	No	No
5A	No	No
5Ba	No	No
5	No	No

this procedure has three tasks, the event tree has a dedicated success and failure line for each task.

(Step 3) Identify negative outcome(s)

Three outcomes can be water leakage into the reheater. However, the failure of Task #3 alone would exacerbate a leak. If the drain valves are not opened, then if the block valves leak, water could accumulate in the reheater and piping prior to startup. Therefore, rename Tasks #1 and #2 as Tasks #1a and #2 and consider them a sub-procedure. Rename Task #3 as Task #1b and consider it another sub-procedure.

(Step 4) For the tasks on the Event Tree, use Checklist A

The answers are listed under Task #a and Task #b in Table D.4b

(Step 5) From Checklist A answers, use Checklist B to assign HEPs

Given the answers to the questions in Checklist A for Tasks #1a and #2a in one sub-procedure and Task #1b in the other sub-procedure, go to Checklist B, read the probability and write it on the Task failure line.

(Step 6) Calculate the HEP for a particular outcome in the procedure

The human error probability for Tasks #1a and #2a in the 1st sub-procedure is 0.03 and the human error probability for Task #1b in the 2nd sub-procedure is 0.09. When these are combined, the human error probability for the procedure is 0.09.

(Step 7) State findings and recommendations

Checklists A and B may not represent an entire procedure. In this case, after drawing the event tree, the procedure was divided into two sub-procedures.

Using Checklist B, here are some ways to reduce the HEP from 0.09:

- Verifying with a checklist that the drain valves are open will reduce the HEP to 0.01.
- Testing that the drain valves are open will reduce the HEP to 0.001. Adding a checklist to verify the status immediately after a test does NOT further decrease the HEP.

Active HEP 2

(Step 1): Translate the written procedure to a timeline

The procedure for Active HEP 1 has two tasks:

Task 1—to detect flow in water line.
Task 2—to open the drain valve.

(Step 2) Translate the tasks into an Event Tree

The tasks in Step 1 are shown as an event tree in Figure D.12a. Because this procedure has two tasks, the event tree has a dedicated success and failure line for each task.

(Step 3) Identify negative outcome(s)

Two outcomes can be leaks. For Task #1, if flow is not detected, water will leak. For Task #2, if water is detected but the operator does not open the drain valves, water will leak. Therefore, renumber Task #1 as Task #1a and consider it a sub-procedure. Rename Task #2 as Task #1b and consider it another sub-procedure.

(Step 4) For each task on the event tree, use Checklist A

The answers are listed beneath Task #a and Task #b in Table D.5

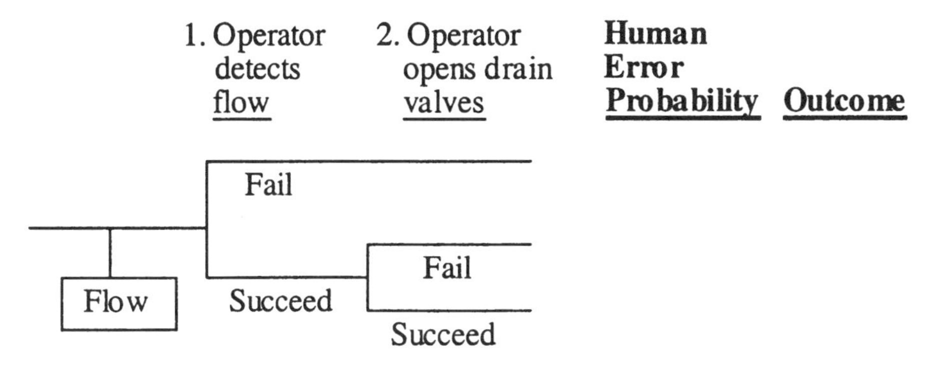

Figure D.12a Reheat Attemperator Water Supply, Active HEP 2, Step 2

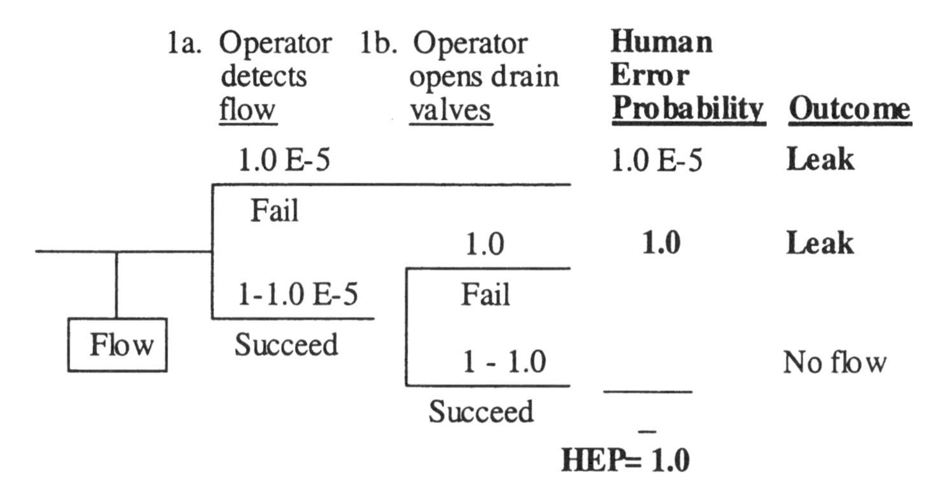

Figure D.12b Reheat Attemperator Water Supply, Active HEP 2, Steps 3 to 6

(Step 5) From Checklist "A" answers, use Checklist "B" to assign HEPs

Given the answers to the questions in Checklist A for Task #1 in one sub-procedure and Task #1 in the other sub-procedure, go to Checklist B, read the probability and write it on the Task failure line. However, because the flow alarm has been a nuisance alarm ("often" indicates flow when there is none) the operators ignore it.

(Step 6) Calculate the HEP for a particular outcome in the procedure

The human error probability for Task #1 in the 1st sub-procedure is 1.0 E-5 but the human error probability for Task #1 in the 2nd sub-procedure is 1.0 rather than 0.05 because FIA is usually a nuisance. When these are combined, the human error probability for the procedure is 1.0.

TABLE D.5 REHEAT ATTEMPERATOR WATER SUPPLY, ACTIVE HEP 2, STEP 4

Question	Task #a	Task #b
3A	1	2
(reason)	*one task*	*one task for both drain values*
3Ba	series	series
(reason)	*water leaks if operator fails to detect flow*	*water leaks if operator fails to open both valves*
3B	Zero	Zero
2	Yes	Yes
1	Yes	No
(reason)	*FIA is feedback*	*cannot see valve position*
4	–	No
5Aa	–	No
5Ab	–	No
5A	–	No
5Ba	–	No
5	–	No

(Step 7) State findings and recommendations

Checklists A and B may not represent an entire procedure. In this case, after drawing the event tree, the procedure was divided into two sub-procedures.

Using Checklist B, here are some ways to reduce the HEP from 1.0:

- Consider a policy that replaces or fixes critical instruments that are a nuisance.

D.4.2 □ Human Error(s) (Constant Rate) As a Single Component

The next two examples discuss the human error probabilities that are associated with two types of inspections. One example involves a visual inspection of a steam drum and the other involves an ultrasonic angle-beam inspection of a weld for preventive maintenance. Both examples have human error as the one "component" that fails.

CASE 3: Visual inspection of steam drum internals

This section assesses the human error contribution to the risk associated with a steam drum inspection. The assessment is divided into Preliminary and Human Error Assessment steps.

Preliminary Steps

The preliminary steps help you to find where human error will contribute to system failure.

(Step 1) Describe the system

A steam drum inspection procedure contains these seven elements:

1. Inspect for accumulated sludge or scale.
2. Inspect cans and scrubber for loose, missing or corroded parts.
3. Inspect scrubber elements for a tight fit.
4. Inspect belly plates for loss of attachments.
5. Inspect drum plates for corrosion, cracks and pitting.
6. Inspect can connections to plates for a tight fit.
7. Inspect feedlines for corrosion and ensure that attachments are tight.

However, this section will assess only Components 2 and 7 for human error.

(Step 2) Describe the consequence(s)

Scenario

A component inside the steam drum is about to fail and the visual inspection of that component fails to detect the incipient failure.

Consequence(s) and severity

Steam drum efficiency will decrease.

(Step 3) Translate the scenario into a fault tree to find human error opportunities

a. For this scenario, a fault tree does not need to be developed to find the basic event(s). For this example, the failed inspection is the human error.
b. For this human error basic event, identify the tasks in a procedure, use Checklist A to identify the human factors involved in the task, and then use Checklist B to assign a human error probability. This step is fully developed in the next section.
c. This step does not apply because there is no fault tree to assess.

Human Error Assessment Steps

This scenario contains one procedure that is a human error opportunity and it is latent. It contributes to a delayed consequence; the error is undetected for a long time (e.g. hours, days, months) until the failing component actually fails.

(Step 1) Translate the procedure to a timeline

Because the procedures for Tasks #2 and #7 are simple, the timeline for each task in these procedures is a single point. Therefore, the two tasks are merely listed:

Task #2 = Visually inspect cans and scrubber for loose, missing, or corroded parts.

Task #7 = Visually inspect feedlines for corrosion and to ensure that the attachments are tight.

(Step 2) Translate the tasks into an event tree

The tasks in Step 1 are shown as event trees in Figure D.13. Because each procedure has only one task, the event tree has a dedicated line for success and failure for that task.

(Step 3) Identify negative outcome(s)

For Task #2, if loose, missing, or corroded parts are not identified, when the boiler restarts, it may operate at a lower efficiency. For Task #7, if corroded feedlines are not identified, then water chemistry could be upset (e.g. too much oxygen), causing, e.g., accelerated corrosion and/or reduced efficiency.

(Step 4) For the tasks on the event tree, use Checklist A

The expected answers are listed as Task #2 and Task #7 in Table D.6.

(Step 5) From Checklist A answers, use Checklist B to assign HEPs

For each task, given the answers to the questions in Checklist A, go to Checklist B, read the probability and write it on the task failure line. See Figure D.13d.

(Step 6) Calculate the HEP for a particular outcome in the procedure

Because each task is the procedure, the human error probability for procedure #2 is 0.05 and for procedure #7 is 0.00005.

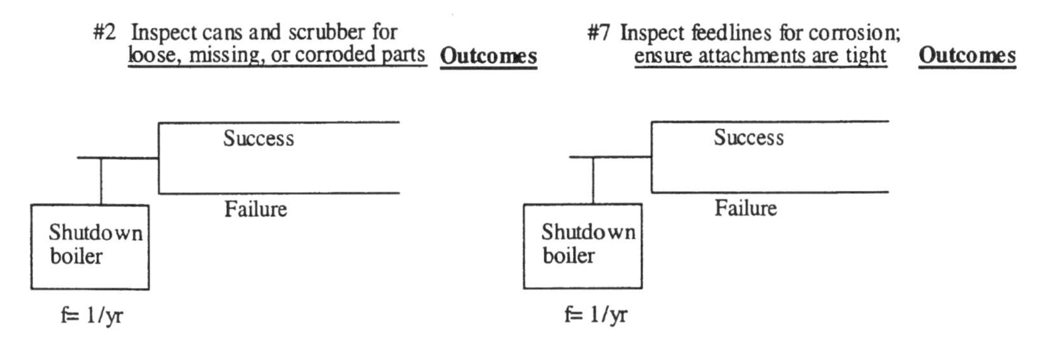

Figure D.13a Visual Inspection Example, Step 2

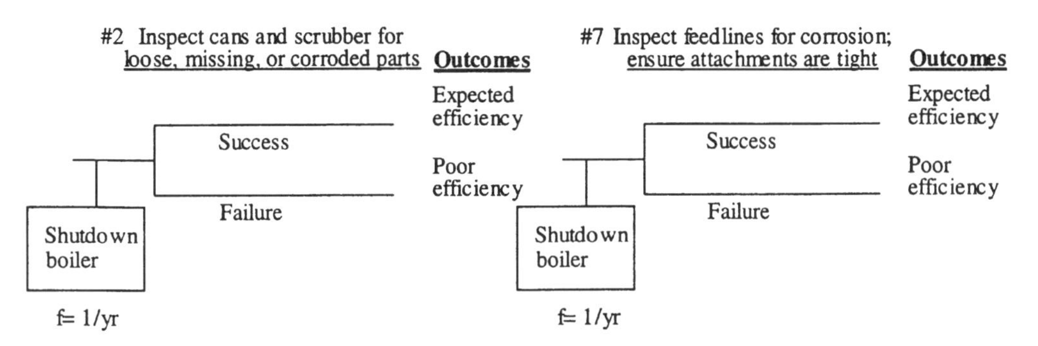

Figure D.13b Visual Inspection Example, Step 3

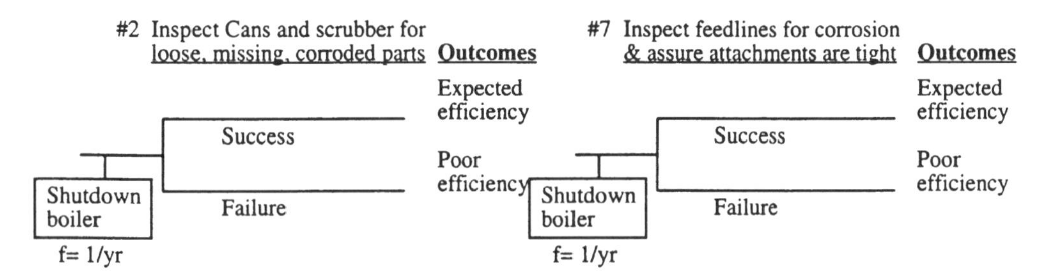

Figure D.13c Visual Inspection Example, Step 4

(Step 7) State findings and recommendations

The probability of failure to detect something missing in Task #2 is greater than that of failing to detect corrosion in Task #7. Task #2 has no feedback, whereas Task #7 has feedback because rust contrasts with a typical pipe color. (The probability assigned to Task #7 assumed that the lines are unobstructed and accessible, with good lighting provided.)

To reduce the human error probability for Task #2 from 0.05 to 0.008,

TABLE D.5 VISUAL INSPECTION EXAMPLE, STEP 4

Question	Task #2	Task #7
3A	1	1
(reason)	*task is the inspection*	*task is the inspection*
3Ba	series	series
(reason)	*poor efficiency if inspection fails*	*poor efficiency if inspection fails*
3B	Zero	Zero
2	Yes	Yes
1	No	Yes
(reason)	*overlook missing parts*	*can see rust*
4	No	—
5Aa	No	—
5Ab	No	—
5A	No	—
5Ba	No	—
5	No	—

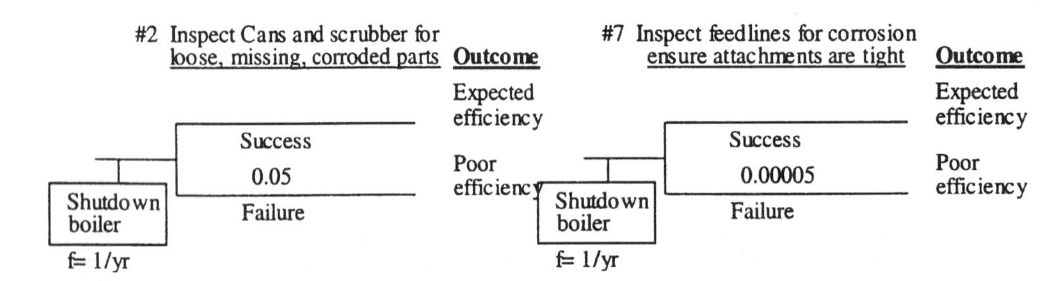

Figure D.13d Visual Inspection Example, Step 5

consider verifying the inspection with a checklist or have another person re-inspect the steam drum.

CASE 4: Inspect a weld

This section assesses human error contributing to a risk with a weld. Note that a complete analysis would need to consider other probability of detection (POD) issues. The assessment is divided into Preliminary and Human Error Assessment Steps.

Preliminary Steps

Preliminary steps help you to find the opportunities where human error will contribute to failing the system.

(Step 1) Describe the system

Several inspection techniques are often used during preventive maintenance inspections to reduce downtime. This section will assess, for human error, an ultrasonic angle-beam inspection of a pipe weld for discontinuities.

(Step 2) Describe the consequence(s)

Scenario

A pipe is welded incorrectly and the UT inspection fails.

Consequence(s) and severity

A failed component may leak or rupture after unit startup and require a shutdown for repair. Economic consequences are certain and safety consequences are possible.

(Step 3) Translate the scenario into a fault tree to find human error opportunities

a. For this scenario, a fault tree does not need to be developed to find the basic event(s). For this example, the failed inspection is the human error.
b. For this human error basic event, identify the tasks in a procedure, use Checklist A to identify the human factors involved in the task, and then use Checklist B to assign a human error probability. This step is fully developed in the next section.
c. This step does not apply because there is no fault tree to assess.

Human Error Assessment Steps

(Step 1) Translate the procedure to a timeline

This timeline shows the technique (adopted from ASME Code Section V, Article 5, T-543) for angle-beam ultrasonic inspection (UT).

1. Calibration
   - Use additional blocks when appropriate.
   - To determine angle beam exit point, use IIW block to check beam angle and distance calibration. (Use miniature angle beam calibration block for 0.25 in. and 5 in. dia. transducers)
   - Examine at gain setting at least 2x reference level (6 dB above)

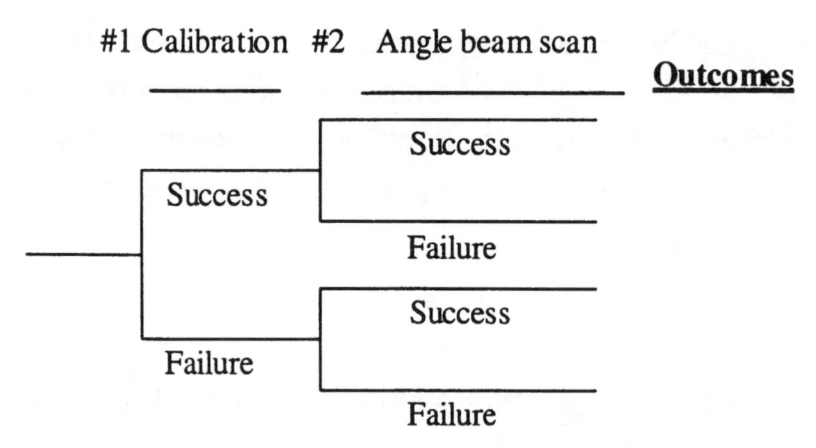

Figure D.14a UT Inspection Example, Step 2

2. Angle beam scanning with beam axis at about 90 degrees to the weld axis from both sides of the weld (whenever possible). Manipulate the search unit so ultrasonic energy passes through the required volumes of weld and adjacent base metal.
3. Angle beam scanning with unit parallel to the weld axis.
4. Longitudinal beam scanning supplements angle beam scanning and/or when geometric considerations make it impossible to cover the required volume from all angles by angle beam procedures and picks up defects parallel to the surface.

(Step 2) Translate the tasks into an event tree

The Angle Beam procedure in Step 1 is shown as an event tree below. Because the procedure has two tasks, each branch of the event tree has a dedicated line for success and failure. See Figure D.14a. Recall that the probabilities that are shown in #2 do not include the POD that is inherent in the procedure that is being used.

(Step 3) Identify negative outcome(s)

If either Task #1 or Task #2 fails, a discontinuity in a weld may allow leakage (or rupture) after the unit starts operating. Unit shutdown will be required to repair the leak. See Figure D.14b.

(Step 4) For the tasks, on the Event Tree, use Checklist A

The expected answers are listed beneath each task in Table D.6.

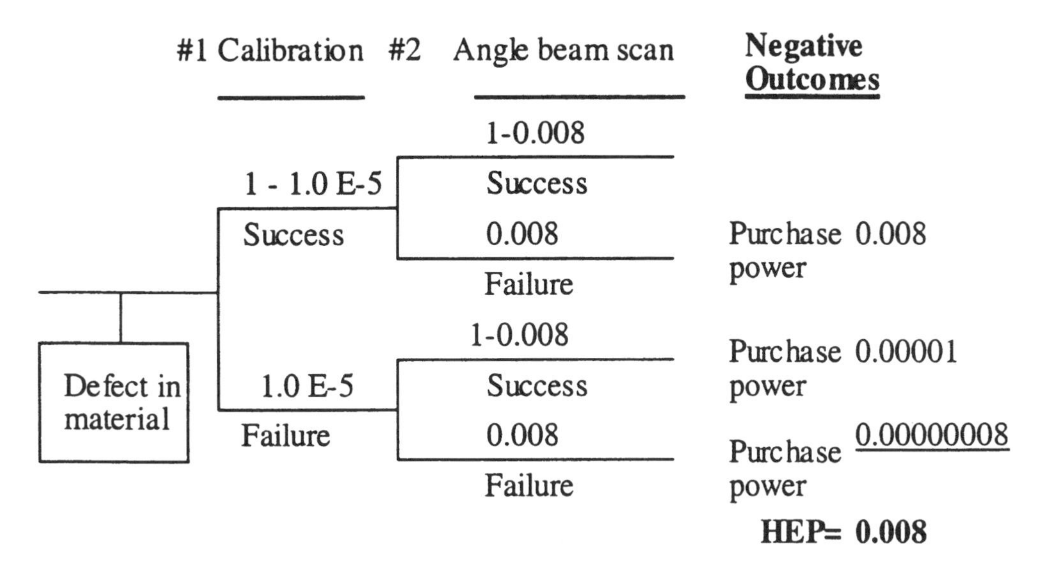

Figure D.14b UT Inspection Example, Steps 3 through 6

(Step 5) From Checklist A answers, use Checklist B to assign HEPs

For each task, given the answers to the questions in Checklist A, go to Checklist B, read the probability and write it on the task failure line. The success value for a given task is 1 minus all the failure values for the task. Figure D.14b shows the probabilities.

(Step 6) Calculate the HEP for a particular outcome in the procedure

To calculate the human error probability (HEP) for the procedure, multiply the human error and success probabilities for Task #1 times the human error and success probabilities for Task #2 and then add the human error probabilities for each negative outcome.

In this example, to get probabilities for the three negative outcomes:

1. Multiply Task #1 success (1-1.0E-5) by Task #2 failure (0.008) to get 0.008
2. Multiply Task #1 failure (1.0 E-5) by Task #2 success (1-0.008) to get 1E-5
3. Multiply Task #1 failure (1.0 E-5) by Task #2 failure (0.008) to get 8E-8
4. Add negatives outcomes (8.0 E-3, 1.0 E-5, 8.0 E-8) to get HEP equal to 8.0 E-3

TABLE D.6 UT INSPECTION EXAMPLE, STEP 4

Question	Task #1	Task #2
3A	2	2
(reason)	*calibration has 2 tasks*	*90°, parallel, longitudinal angle beam scan*
3Ba	parallel	parallel
(reason)	*both tasks need to fail*	*all tasks need to fail*
3Bb	<2 minutes	>2 minutes
3Bc	<4 feet	
3B	Complete	Zero
2	Yes	Yes
1	Yes	No
(reason)	*comparison*	–
4	–	No
5Aa	–	No
5Ab	–	Yes
5Ac	–	Yes
5A	–	Yes
5Ba	–	No
5B	–	No

(Step 7) State findings and recommendations

The probability of failing to detect a flaw in Task #1 is less than that of failing to detect a flaw in Task #2. Task #1 has no compelling feedback whereas Task #2 has verification feedback (when the longitudinal scan checks the angle beam scan). The task #2 assessment assumes that the weld is unobstructed and accessible and that the area is well-lighted.

Therefore,

- No need to reduce the human error probability of Task #1 from 1.0 E-5.
- To reduce the human error probability of Task #2 from 0.008 to 0.0008, consider verifying the inspection with a checklist or have another person re-inspect the weld. (In practice, such welds are generally hydrostatic tested.)

Recall once more that a complete analysis also requires you to consider the POD that is associated with the inspection technique. In this case, the POD could increase the failure probability by as much as two orders of magnitude.

D.A □ CHECKLIST A

Human Factors in ASEP Normal Operation

INSTRUCTIONS: Answer each of the following 6 questions. Transfer your ratings to CHECKLIST B to identify the mean human error probability for the tasks in a procedure.

ASEP TASK: ________________
FACILITY: ________________
ANALYSTS: ________________
DATE: ________________

1. COMPELLING FEEDBACK (You know immediately to correct an error)
Is the wrong action fully corrected because overwhelming and fast feedback, such as signals from one or more annunciators, alarms, valve position indicators, or anything else, can lead the operator to correct the error after the task has been completed but before it leads to an undesired outcome?

1. Is the feedback compelling?
YES **(1)**
NO

2. ADEQUATE ADJUSTORS (Procedures, training, man/machine interfaces)
If overall attention to administrative controls, emergency and operating procedures, and training is very poor or if the human-machine interface is quite deficient rate adequacy as NO, otherwise rate as YES.

2. Are adjustors adequate?
YES
NO

3B. RATE DEPENDENCY AMONG TASKS Order of questions ------- >

3A. Number of tasks in the procedure:______

a) Tasks are within the context of which type of system?	b) How much time is there between tasks on similar components?	c) For tasks on similar components, how close together are the components?	d) Written sign-offs are made for each task?	**3B. Rate the task dependency**
Parallel (AND) (need 2 or more tasks to fail before the system fails)	≤ 2 minutes	Separated ≤ 4 feet		COMPLETE
		Separated > 4 ft but same general area	Yes	ZERO
			No	HIGH
		Different areas		ZERO
	> 2 minutes			ZERO
Series (OR) (just 1 failed task among 1 or more tasks will fail the system)				ZERO

4. OBVIOUS FEEDBACK (test the system **P**ost-**M**aintenance or **P**ost-**C**alibration)
Can maintenance or calibration err(s) be identified and corrected by correctly testing the system before returning it to service

4. PM/PC tested?
YES
NO

Human Factors in ASEP Normal Operation

Checklist A (Continued)

5A. IMMEDIATE SUBTLE FEEDBACK (Verify task(s) result) Order of questions ---- >

a) 2nd individual personally verifies task(s) result ? **(2)**	b) Performer checks task(s) result at different time and place ?	c) Is a written check-off used ?	**5A. Is the result verified ?**
No	Yes	Yes	YES
		No	NO
	No		NO
Yes		Yes	YES
		No	NO

5B. PERIODIC SUBTLE FEEDBACK (Check status of task(s) result regularly) Order of questions --->

a) Is status checked regularly per batch or shift ?	b) Is a written check-off used ?	**5B. Is the status checked ?**
Yes	Yes	YES
	No	NO
No		NO

NOTE 1: If there is a compelling signal, then no further ratings are needed.

NOTE 2: Answer Yes only if the second individual personally verifies the results of the task(s).

NOTE 3: This job aid corresponds completely to the ASEP Nominal HRA for pre-methodology as set forth in NUREG/CR 4772 Chapter 5

D.A □ CHECKLIST B

Human Error Probability - ASEP Normal Operation

INSTRUCTIONS: Transfer your ratings from CHECKLIST A to identify the mean human error probability for the tasks in a procedure.

ASEP TASK: ______________________

FACILITY: ______________________

ANALYSTS: ______________________

DATE: ______________________

1. **If Compelling Feedback = Yes then HEP = Negligible** (Upper Bound < 0.00001)

2. Adjustor	**3B.** Depend	**4.** Obvious or Subtle Feedbacks	**5A.**	**5B.**		**Mean HEP** (Error Factor) **3A.** Number of tasks in a procedure				
Train, instruct	Tasks relate	PM/PC tests	Veri-fied	Status Checks	Row ID	1	2	3	4	5
Ade-quate	Zero	Yes	Yes	Yes	1	.0001 (16)	.0002 (9)	.0002 (7)	.0002 (6)	.0002 (6)
				No	2	.0008 (10)	.001 (5)	.001 (4)	.002 (4)	.002 (3)
			No	Yes	3	.0001 (16)	.0002 (9)	.0002 (7)	.0002 (6)	.0003 (6)
				No	4	.0008 (10)	.001 (5)	.001 (4)	.002 (4)	.002 (3)
		No	Yes	Yes	5	.0008 (10)	.001 (5)	.001 (4)	.002 (4)	.002 (3)
				No	6	.008 (10)	.01 (7)	.02 (6)	.02 (5)	.02 (4)
			No	Yes	7	.008 (10)	.01 (7)	.02 (6)	.02 (5)	.02 (4)
				No	8	.05 (5)	.09 (4)	.11 (3)	.15 (3)	.16 (2)
	High	Yes	Yes	Yes	9	-	<<<	<<<	<<<	<<<
				No	10	-	.0002 (8)	.0001 (9)	.0001 (10)	<<<
			No	Yes	11	-	<<<	<<<	<<<	<<<
				No	12	-	.0002 (8)	.0001 (9)	.0001 (10)	<<<
		No	Yes	Yes	13	-	.0002 (8)	.0001 (9)	.0001 (10)	<<<
				No	14	-	.003 (11)	.002 (12)	.0008 (13)	.0001 (14)
			No	Yes	15	-	.003 (11)	.002 (12)	.0008 (13)	.0001 (14)
				No	16	-	.02 (6)	.01 (7)	.005 (7)	.003 (8)
	Comp-lete	Yes	Yes	Yes	17	-	.0001 (16)			
				No	18	-	.0005 (10)			
			No	Yes	19	-	.0001 (16)			
				No	20	-	.0005 (10)			
		No	Yes	Yes	21	-	.0005 (10)			
				No	22	-	.005 (10)			
			No	Yes	23	-	.005 (10)			
				No	24	-	.03 (5)			

APPENDIX D: HUMAN FACTORS

Human Error Probability - ASEP Normal Operation

Checklist B (Continued)

1. If Compelling Feedback = Yes then HEP = Negligible (Upper Bound < 0.00001)

2. Adjustor	3B. Depend	4. Obvious or Subtle Feedbacks	5A.	5B.		Mean HEP (Error Factor) 3A. Number of tasks in a procedure				
Train, instruct	Tasks relate	PM/PC tests	Veri-fied	Status Checks	Row ID	1	2	3	4	5
Inade-quate	Zero	Yes	Yes	Yes	25	.0002 (16)	**.0002 (9)**	.0003 (7)	.0004 (6)	.0005 (6)
				No	26	.001 (10)	**.001 (5)**	**.001 (4)**	**.002 (4)**	**.002** (3)
			No	Yes	27	.0002 (16)	**.0002 (9)**	.0003 (7)	.0004 (6)	.0005 (6)
				No	28	.001 (10)	**.001 (5)**	**.001 (4)**	**.002 (4)**	**.002** (4)
		No	Yes	Yes	29	.001 (10)	**.001 (5)**	**.001 (4)**	**.002 (4)**	**.002** (4)
				No	30	.01 (10)	.02 (7)	.03 (6)	.03 (5)	.04 (4)
			No	Yes	31	.01 (10)	.02 (7)	.03 (6)	.03 (5)	.04 (4)
				No	32	.08 (5)	.14 (4)	.19 (3)	.25 (3)	.27 (2)
	High	Yes	Yes	Yes	33	-	.0001 (14)	<<<	<<<	<<<
				No	34	-	**.0002 (8)**	**.0001 (9)**	**.0001** (7)	<<<
			No	Yes	35	-	<<<	<<<	<<<	<<<
				No	36	-	**.0002 (8)**	**.0001 (9)**	**.0001** (7)	<<<
		No	Yes	Yes	37	-	**.0002 (8)**	**.0001 (9)**	**.0001** (7)	<<<
				No	38	-	.005 (11)	.003 (12)	.001 (13)	.0008 (14)
			No	Yes	39	-	.005 (11)	.003 (12)	.001 (13)	.0008 (14)
				No	40	-	.03 (6)	.02 (7)	.008 (7)	.005 (8)
	Comp-lete	Yes	Yes	Yes	41	-	**.0001 (16)**			
				No	42	-	.0009 (10)			
			No	Yes	43	-	**.0001 (16)**			
				No	44	-	.0009 (10)			
		No	Yes	Yes	45	-	.0009 (10)			
				No	46	-	.009 (10)			
			No	Yes	47	-	.009 (10)			
				No	48	-	.05 (5)			

NOTES: <<< = Negligible (Mean HEPs were less than 5.0 E-5)
Mean HEPs between 5.0 E-5 and 9.0 E-5 were rounded up to 1.0 E-4

APPENDIX E
INDEX TO NERC-GADS SYSTEM/ COMPONENT CAUSE CODES

This appendix lists the NERC-GADS cause codes and, in lightly edited form, the instructions that should have been followed when they were assigned by the persons who submitted them. The appendix itself is APP_E.pdf, which is located on the Handbook CD-ROM in the /C_CODES subdirectory. You will need Adobe Acrobat Reader, which is located elsewhere on the CD-ROM, to use it. The appendix contents are as follows:

APPENDIX

HOW TO GET HELP

The Handbook comes with an offer of assistance for questions both general and specific. If the following FAQs don't provide the information you need, please e-mail your inquiry to **riskhandbook@asme.org**

FAQS: General Content

Q. *What is the purpose of the Handbook?*

A. The Handbook contains step-by-step procedures that should help its users to properly pose equipment remaining life analysis and management questions and to apply previously-developed, in fact classical, analytical tools to help answer those questions.

Q. *Who is the intended audience?*

A. Appropriate users of the Handbook hold two qualifications:

- Technical expertise—as the Handbook states early and often, its audience is engineers who are expert in the facilities and processes to which they will apply it. The Handbook makes no attempt to exhaustively inform its users about the hazards of anything; rather, it explains ways to approach certain sample hazards and directs its users to apply their own resources to extend the examples.
- Computer literacy—although many of the Handbook-recommended tools predate the computer, their practical application, specifically as suggested by the Handbook and illustrated by its examples, requires at least minimal comfort with a personal computer and with spreadsheet software.

Q. *What is "remaining life analysis and management"?*

A. "Remaining life analysis and management" involve predicting the remaining life, in this case, of plant components, and making repair and replacement decisions about them. Such components' lives are "consumed" by damage mechanisms that result from load, temperature, environment or a combination of these. The Handbook stresses that the analysis method(s) used is(are) probabilistic. This is because the remaining life prediction is uncertain.

Q. *What is inherently uncertain about remaining life prediction?*

A. It is rarely possible and probably never practical to know all the aging mechanisms that are occurring in a component and how far the mechanisms have advanced in a specific component in service. The only statement that anyone can make with 100% certainty is that every component will some day fail, probability of 1. This is not a useful conclusion, so the analysis that the Handbook recommends expresses remaining life as a cumulative probability of failure versus time on a scale from 0 to 1. This explicitly recognizes that the estimated time of failure is uncertain. Further, the probability is expressed as the increasing probability of failure over time and not as the certainty of failure at any moment in time. Expressing it this way also makes it unlikely that a Handbook user will interpret the analysis results as a "definite" predicted time of failure.

Q. *What could possibly go wrong if the tools were misused? An airplane crash? Explosion of a power generator?*

A. If a prediction model was applied to the wrong sort of component, e.g., one that is subject to failure mechanism(s) that are not addressed by the model (something against which users are specifically cautioned in the Handbook), then the resulting program(s) might calculate an inappropriately low failure probability. This could create sense of security that could prove false.

The user could also use an appropriate prediction model but the wrong optimizer. When safety is an issue, the user is directed to a model that explicitly limits the "allowed" failure probability that the user is advised to obtain from management.

Q. *Does a wrong model mean that a failure will result?*

A. We need to emphasize that using the wrong model does not necessarily mean that a failure will occur. The worst-case safety-related component that the Handbook considers by example is high-energy steam piping. Such piping has, prior to its failure mechanism being fully understood and inspections focused on preventing it, ruptured 4 times in the last 10 years, with only one incident resulting in multiple fatalities. In this case, the hazard applies to plant workers, not the public, because failures stay within the plant boundary, which has controlled access. Even this "base" level of incident frequency has not resulted in a worker fatality probability that is any greater than that caused by driving a car to work. Users with more serious hazards are advised, through management consultation, to apply even more restrictive safety criteria.

Q. *What precautions should be taken by a user who has a potential safety exposure?*

A. Know the facility, know its hazards, and get reality-checked every step of the way. For example, the safety-related (high-energy piping) model that is used in the Handbook was written by a metallurgist in a medium-sized electric utility. It was reviewed by one of the top industry experts. This person wrote the generally-accepted prediction model (for the Electric Power Research Institute) that is distributed to utilities for probabilistically predicting the remaining life of high pressure components like piping.

Q. *What about other industries?*

A. The Handbook could also be applied to other industries, for which similar cautions apply and are stated. No industry could tolerate frequent, catastrophic failures. This suggests that reasonably good models could, with care, be developed for most components. On the other hand, in the utility industry or any other industry, a previously-unknown failure mechanism could cause predictions to be less accurate. In all cases, though, a probability curve has a tail. Some failures will occur no matter how sound the analysis was that said that they wouldn't. (The layperson might call this "bad luck.") Finally, we argue quite forcefully that the probabilistic approach is fundamentally less dangerous than any approach that implies that it reveals some sort of precise truth.

FAQs: CD-ROM

Q: *What are the platform requirements?*

A: All of the programs on the CD-ROM, except as noted below, are designed to run on any PC that is running Microsoft Windows 98 or later. Platform requirements specific to the various programs (which are listed in the table "CD-ROM Contents" located later in this appendix) are:

1. All files with .pdf extensions require Adobe Acrobat or Adobe Acrobat Reader, version 3.0 or higher. (Adobe Acrobat Reader is included on the CD-ROM.)
2. All files with .xls extensions require Microsoft Excel 8 (the version in Microsoft Office 97) or later, however, the Mantop*.xls files may have a minor issue with Excel 9/Office 2000 on some machines. See the readme file in the \EXAMPLES subdirectory for more information and the workaround.
3. The Mantop*.xls files require the Excel Solver add-in.
4. Pipesafe.xls, Tubemod.xls, Tubestra.xls and Tubstrsk.xls require Palisade @Risk, a commercial spreadsheet add-in, version 4.5. @Risk requires an Intel Pentium or higher microprocessor (or the equivalent) and 8 MB installed RAM (16 MB for Windows NT). See the readme file in the \@RISK subdirectory for more information. (A trial version of @Risk is included on the CD-ROM.)
5. IRRAS requires DOS 3.3 or higher, any of the DOS-based Windows versions or OS/2. See the readme file in the \IRRAS subdirectory for more information.

Q: *What's on the CD-ROM?*

A: This table lists the "parent" applications for the CD-ROM. It does not, for example, list the 100 or so files in the IRRAS package or the .pdf index files.

CD-ROM Contents

File or Group	Description	Data
RBMap.pdf	A program development process flowchart that is hot-linked to digital copies of the Handbook instructions	A subset of the sample data that is printed in the book
App_E.pdf	A digital version of the NERC-GADS cause codes	Equipment lists with assigned codes
FTHB.pdf	A digital version of the *Fault Tree Handbook*, an out of print NUREG	None
\IRRAS	Files for the Integrated Reliability and Risk Analysis Software	–
Reference.pdf	A digital version of the IRRAS Reference Manual, an out of print NUREG	None
Tutorial.pdf	A digital version of the IRRAS Tutorial, an out of print NUREG	None
.\Program	The IRRAS program files	Demo data from the authors
.\Sample	A fault tree for a fictitious power plant	Same example as in the Handbook and Volume 3
\EXAMPLE	Spreadsheet templates for Handbook examples	–
Baycom11.xls	Combines probability distribution curves from history and interview	Probcalc.xls and Probcaus.xls
Baycom12.xls	Combines probability distribution curves from Baycom11 and an engineering model	Baycom11.xls and PipeSafe.xls or Tubemod.xls
Conseq.xls	Calculates total probability and total consequence given failure probability curves	Baycom11.xls and Baycom12.xls
Mantop.xls	Financially based component replacement timing	Baycom11.xls, Baycom12.xls and Tubemod.xls
mantops.xls	Safety constrained financially based component replacement timing	Baycom11.xls, Baycom12.xls and PipeSafe.xls
PipeSafe.xls	Probabilistic remaining ilfe analysis model for cracked high pressure containing components	Synthetic data
Probcalc.xls	Calculate the change in probability by year	Synthetic data
Probcaus.xls	Calculate the change in probability by year	NERC-GADS

File or Group	Description	Data
Riskcaus.xls	Combines failure history consequence data by component to produce elements of risk	NERC-GADS
Riskplot.xls	Graphically plots the elements of risk using production loss	Riskcaus.xls and Riskrank.xls
Riskplt$.xls	Graphically plots the elements of risk using dollar loss	Conseq.xls
Riskrank.xls	Combines failure history consequence data by component to produce elements of risk	Synthetic data
Tire.xls	Simple strategy table building example	Synthetic data
Tubemod.xls	Boiler tube probabilistic remaining life analysis	Synthetic data
Tubestra.xls	Simple strategy table building example	Synthetic data
Tubstrsk.xls	Simple strategy table building example with sensitivity analysis	Synthetic data
\READER	Copies of Acrobat Reader for various platforms	None

Q: *How will the reader actually use the CD-ROM?*

A. The reader will take plant component geometry and operation data and inspection results for the selected component(s) and insert them in the probabilistic model on the CD-ROM in place of the demo data that will already be there. Running the program will produce a probability of failure over time curve. This result will be inserted into a component replacement timing financial model (also on the CD-ROM) that converts the probability of failure year by year into standard financial calculations that provide the most optimum financial time to replace the component based on the combination of the probability versus time and the component replacement cost. The reader is reminded that the optimum value created is an expected value (rather like an average), not an exact value.

Q: *May the disk be installed on a network drive?*

A: Yes, but for multiple users, a site license is required. Email ASME at permissions@asme.org for more information.

Q: *May any of the CD-Rom contents be modified?*

A: No, except for the following:

1. The IRRAS examples may be modified at will.
2. The data entry cells only of the .xls (spreadsheet) files in the /EXAMPLES subdirectory may be modified as directed in the applicable procedures. (The data entry cells are yellow colored, however, they may appear pale orange on some monitors.)

NOTES:

NOTES:

NOTES:

NOTES:

NOTES: